T0295508

BIOCHEMISTRY RESEARCH TRENDS

BIOLUMINESCENCE

CHARACTERISTICS, ADAPTATIONS AND BIOTECHNOLOGY

BIOCHEMISTRY RESEARCH TRENDS

Additional books in this series can be found on Nova's website under the Series tab.

Additional E-books in this series can be found on Nova's website under the E-book tab.

BIOCHEMISTRY RESEARCH TRENDS

BIOLUMINESCENCE

CHARACTERISTICS, ADAPTATIONS AND BIOTECHNOLOGY

DAVID J. RODGERSON
EDITOR

Nova Science Publishers, Inc.
New York

For permission to use material from this book please contact us:
Telephone 631-231-7269; Fax 631-231-8175
Web Site: http://www.novapublishers.com

Library of Congress Cataloging-in-Publication Data

Bioluminescence : characteristics, adaptations, and biotechnology / editor, David J. Rodgerson.
p. ; cm.
Includes bibliographical references and index.
ISBN 978-1-61209-747-3 (hardcover)
1. Bioluminescence. I. Rodgerson, David J.
[DNLM: 1. Luminescence. 2. Nanotechnology--methods. QH 641]
QH641.B557 2011
572'.4358--dc23

2011026363

Published by Nova Science Publishers, Inc. † New York

CONTENTS

PREFACE

This new book presents current research data in the study of the characteristics, adaptations and biotechnology of bioluminescence. Topics discussed include ATP-Driven bioluminescence; the coupling of luciferases to nanomaterials; the characteristics and adaptations of bioluminescent bacterial biosensors for acyl-HSL signal detection, quantification and tentative identification of acyl-HSLs; using bacterial biosensors to determine the toxicity of nanomaterials and other xenobiotics and the visual spectral mechanisms and bioluminescent optical signaling in beetles.

Chapter 1 - Dental caries is a very common infectious disease primarily transferred to children during early childhood from maternal and other close-contact sources. The principal cariogenic microorganisms are highly acidogenic and acid-tolerant bacteria and include members of the mutans streptococci group and lactobacilli. Numerous assessment tools for oral hygiene and caries risk are available to identify patients at risk for dental caries, but better quantitative methods are needed to allow appropriate targeting of aggressive, caries-protective treatments. Rapid adenosine triphosphate-driven (ATP-driven) bioluminescence assays have long been used to quantitatively measure microbial numbers in a variety of situations, including clean room monitoring and in the sanitation and food-processing industries. More recently, ATP-driven bioluminescence has been used to quantitate bacterial load in both medical and dental settings. For example, dental plaque mass and bacterial numbers in both humans and animals correlate well with ATP-driven bioluminescence values. The purpose of this review is to illustrate the use of ATP-driven bioluminescence in the rapid chair-side enumeration of total oral bacteria, including cariogenic plaque streptococci. The authors' studies using oral clinical specimens from pediatric patients demonstrate that ATP-driven bioluminescence can be used in the direct determination of bacterial numbers, and can serve as a general assessment indicator for oral hygiene. ATP-driven bioluminescence may potentially serve as a component of dental caries risk assessment.

Chapter 2 - Luciferase is a powerful tool in bioanalysis. Several well-established methods employ luciferases, particularly firefly and *Renilla* luciferases, as reporter genes or biosensors in environmental, biomedical and biochemical research. These techniques have interesting features for the analyst such as sensitivity, specificity and reduced assay time. Nanochemistry and Nanotechnology are disciplines that are gaining much attention and evolving rapidly. They allow the development of custom-made nanomaterials with the desired properties, starting from conventional bulk materials. Recently, the coupling of nanomaterials such as carbon nanotubes, mesoporous silica nanoparticles, metallic nanoparticles and quantum dots with luciferases led to new or improved methodologies for analyte quantification and enhanced gene delivery strategies. One of the principal scopes is to modulate or alter luciferase's bioluminescence emission, either by stabilizing it or tuning it to longer wavelengths. This chapter aims to present state-of-art articles regarding new methods based on the coupling of luciferases to nanomaterials, along with a brief introduction to Nanoscience.

Chapter 3 - Recombinant whole-cell bioluminescence biosensors developed over the last decade have been very useful to discover bacteria that posses an N-acyl-homoserine lactone (acyl-HSL) regulated quorum-sensing (QS) communication system.The acyl-HSL-dependent bacterial bioreporter systems, also known as specific biosensors were constructed to respond to the presence of a certain small molecules, in this case acyl-HSLs.These acyl-HSL-dependent specific biosensors have been developed becauseacyl-HSL signalsdo not contain strongly absorbing chromophores and acyl-HSLs are difficult to detect with standard chemical tests. In addition, the problem of acyl-HSL detection ischallenging since these signals are produced at very low levels, customarily in the nanomolar to micromolar range. Acyl-HSLbiosensors contain an acyl-HSL signal-dependent receptor / regulator protein that is a member of the LuxR-family and a gene coding for a detectable response phenotype, the reporter gene. The expression of the reporter phenotype occurs only in the presence of acyl-HSL signals and is controlled by anacyl-HSL-LuxR complex responsive promoter fused to the reporter gene(s). A number of bioreporter strains, each containing a different LuxR homolog receptor, have been constructed that report light production or bioluminescence from the *luxCDABE* operon as detection end points. This review aims to provide a background to examine the characteristics and adaptations of bioluminescent bacterialbiosensors for acyl-HSL signal detection, quantification and tentative identification of acyl-HSLs.

Chapter 4 - It is widely accepted that the current and planned increase in commercial development of nanotechnology will lead to pronounced releases of nanomaterials (NM) in the environment. In this study the authors aim to evaluate the toxicity of silver nanoparticles and some chlorinated phenol compounds by using a common sludge bacterium, Pseudomonas putida (originally isolated from activated sludge) as biosensor in bioluminescence assays. The results of the toxicity testing were expressed as IC50 values which represents the amount of toxicants required to reduce light output by 50% and calculated using a statistical program that was developed in-house. The bacterium carrying a stable chromosomal copy of the lux operon (luxCDABE) was able to detect toxicity of ionic and particulate silver over short term incubations ranging from 30-240 min. The IC50 values obtained at different time intervals showed that highest toxicity (lowest IC50) was obtained after 90 min incubation for all toxicants and this is considered the optimum incubation for testing. The data show that ionic silver is most toxic followed by nanosilver particles with microsilver particles being least toxic. Release of nanomaterials is likely to have an effect on the activated sludge process as indicated by this study using a common sludge bacterium involved in biodegradation of organic wastes. Among the chlorinated phenol compounds, 3,5 dichlorophenol showed the highest toxicity and phenol being the least toxic. Among the hydroxylated compounds tested, benzoquinone showed to be the most toxic compound. Understanding the fate and transport of phenolic compounds and transformation products in removal treatment plants is relevant in evaluating the impact of discharge in the environment and in developing a program for pollution control. The authors' results demonstrate that the use of a bacterial biosensor as P. putida BS566::luxCDABE, native from a highly polluted environment, provides a robust, early warning system of acute toxicity which could lead to process failure. This strain is suitable for toxicity monitoring in a highly polluted industrial waste water treatment streams and it has the potential to inform on upstream process changes prior to their impacting upon the biological remediation section.

Chapter 5 - Bioluminescence is a biological phenomenon in which energy is released by a chemical reaction in the form of cold light emission (chemiluminescence). Evolution of bioluminescence has arisen independently as many as 30 times with the five main traits of camouflage, attraction, repulsion, communication and illumination. Bioluminescence is common in bacteria and fungi but is also observed in ctenophores, annelid worms, jelly fish, mollusks, glow worms and fireflies. Bioluminescent symbiotic microorganisms are frequently found in symbiotic association with larger organisms, such as fish. Bioluminescence is caused by pigments produced by the organisms, such as luciferin, coelenterazine, and vargulin. The synthesis of luciferin, the most studied bioluminescent pigment, is catalyzed by the enzyme luciferase that is bound in inactive form to adenosine triphosphate (ATP). Once released from ATP, luciferase causes luciferin to react with molecular oxygen, yielding an electronically excited oxyluciferin species. Relaxation of excited oxyluciferin to its ground state results in emission of visible light. Slight structural differences in luciferase affect both the high quantum yield of the luciferin/luciferase reaction and bioluminescence colour. The luciferin and catalyzing enzyme, as well as a co-factor such as oxygen, are bound together to form a single unit called a "photoprotein". A photoprotein can be triggered to produce light when a particular type of ion (frequently calcium) is added. Bioluminescent pigments can be expressed in recombinant organisms and found numerous applications in biological, medical, and biotechnology research. Luciferases are more sensitive than fluorescent reporters, owing to the extremely low background levels of bioluminescence. Also, there is no need for exogenous illumination, which in fluorescence methods can bleach the reporter, perturbs physiology in light-sensitive tissues (e.g. the retina), and causes phototoxic damage to cells. Therefore, bioluminescence imaging is used to study viable cells, tissues and whole organisms. Molecular imaging techniques represent a revolutionary advancement in our ability to study structural and functional relationships in biology by combining the disciplines of molecular/ cellular biology and imaging technology. Newly developed imaging methods allow transcriptional/translational regulation, signal transduction, protein-protein interaction, oncogenic transformation, cell and protein trafficking, and target drug action to be monitored *in vivo* in real-time with high temporal and spatial resolution, thus providing researchers with priceless information on cellular functions. Those bioluminescent pigments that emit in the visible range can be used for high-throughput screening methods, which are at the forefront of renewable energy applications. Recently, even eukaryotic parasites *Plasmodium, Leishmania* and *Toxoplasma* have been transformed with luciferase and yielded unique insights into their *in vivo* behavior. The applications to biosensing are also interesting and include environmental monitoring of toxic and mutagenic compounds, heavy metals, etc. Other possible applications of engineered bioluminescence include engineering of "smart" organisms and devices, such as glowing trees on highways, novelty bioluminescent pets, bioluminescent agricultural crops and plants at time of watering and bio-identifiers for escaped convicts or mentally ill patients. The discovery of other bioluminescent pigments that emit at different wavelength and the advancements in recombinant technology will in future increase the number of applications.

Chapter 6 - Optical imaging methods such as fluorescence are provided with a broad range of proteins and dyes used to visualize many types of these biological processes widely used in cell biology studies and lately also in clinical research. Although the best-known example of these fluorochromes is the green fluorescent protein (GFP), tissue autofluorescence and signal dispersion raise doubts about its suitability as an *in vivo* tracer. Bioluminescence, on the other hand, does not show this signal attenuation caused by living tissues but relies on a chemical reaction for bioluminescent light to be emitted. Taking into account all of these characteristics of the two most known and used optical methods the authors have devised a novel biosensor comprised by a dual fluorescence-bioluminescence tracer activatable only when the intracellular oxygen concentrations are low enough, or hypoxia. This fusion protein is able to display both fluorescent and bioluminescent properties besides of auto-excitation through a bioluminescent resonance energy transfer or BRET phenomenon, which allows for a better and easier fluorescence performance in localizations where some tissues tend to hinder the fluorochrome excitation.

Chapter 7 - *Staphylocoocus aureus* is an important pathogen and can cause both human and animal infections. The continual emergence of drug resistant, notably methicillin resistant *S. aureus* isolates from hospitalized patients and in the community, as well as in livestock farms has produced a serious public health burden. The availability of a sensitive approach to monitor temporal gene expression or regulation enables us to elucidate molecular mechanism of resistance and pathogenesis. In this study, the authors chose bacterial luciferase (LuxABCDE) as a reporter and created a promoter-*lux* fusion in different *S. aureus* isolates, determined the transcriptional levels of alpha-toxin gene (*hla)* using a luminometer, and found that the temporal expression and the intensity of bioluminescence vary in different isolates carrying the same *hla* promoter-*lux* reporter. The results indicate that the luciferase-driven bioluminescence reporter provides a powerful tool for temporal examination of gene expression and regulation.

Chapter 8 - Bioluminescence resonance energy transfer (BRET) represents a biophysical method to study physical interactions between protein partners in living cells fused to donor and acceptor moieties. It relies on a non-radiative transfer of energy between donor and acceptor, their intermolecular distance (10 – 100 Å) and relative orientation. Several versions of BRET have been developed that use different substrates and/or energy donor/acceptor couples to improve stability and specificity of the BRET signal. In recent years, numerous studies have applied BRET technology to develop screening assays for seven-transmembrane receptors (7TMRs), which represent a key drug target class. In general, these assays are based on 7TMR/β-arrestin interaction, common to virtually all 7TMRs; however, differences in 7TMRs affinity for β-arrestins and the stability/longevity of receptor/β-arrestin complexes exist. This chapter summarizes different approaches (e.g. mutations in β-arrestins, 7TM receptor carboxyl-terminal tail swapping) to optimize the BRET assay for measuring receptor/β-arrestin interaction and applicability of this technological platform for compound medium/high-throughput screening (MHT/HTS).

Chapter 9 - As luminescent bacteria can convert chemical energy into light using luciferase, they are known to glow in the dark with a visible peak wavelength (in the case of *Photobacterium kishitanii*, ca. 475 nm). The organism is thought to have luciferase, evolutionally, to scavenge oxygen. Industrially, luminescent bacteria is used for toxicity measurement. In this system, luminescence intensity is inhibited by the toxic compounds in the sample. Stabilization of the luminescence intensity is therefore important to realize highly sensitive measurement. Experimentally, control of luminescence intensity is not easy. For example, oscillation in the luminescence intensity from the bacterial suspension is often observed in a certain environment. Reaction-diffusion of dissolved oxygen into the cell, and synchronization of luciferase gene expression as a result of quorum sensing, were thought to be two reasons for the oscillation. To explain such a behavior, the characteristics of the cells in suspension was investigated using a microfluidic device. In order to compare the luminescent intensity of bioluminescence from marine luminous bacteria with different motility, luminescent bacteria were separated according to their motility using the device. Calculation of the luminescent intensity per cell was performed, and swimmer cells were shown to be brighter that the others. Luminescence from intact bacteria also show interesting characteristics such as color changing or irradiation-controlled quenching. Understanding of such characteristics will be a key for a novel application. In this chapter, recent findings in the bacterial luminescence is reported, and their application is proposed.

In: Bioluminescence
Editor: David J. Rodgerson, pp. 1-27
ISBN 978-1-61209-747-3

Chapter 1

BIOLUMINESCENCE AND APPLICATIONS IN DENTISTRY

Iraj Kasimi[1,4,*], Christopher W. Kyles[1,4,*], Jill Pollard[1,4,*], Amy Trevor[1,4,*], Alex H. Vo[1,4,*], Tom Maier[1,2], and Curtis A. Machida[1,3,†]

Departments of [1]Integrative Biosciences, [2]Pathology and Radiology and [3]Pediatric Dentistry, and [4]Academic DMD Program, School of Dentistry, Oregon Health & Science University, Portland OR 97239, USA

ABSTRACT

Dental caries is a very common infectious disease primarily transferred to children during early childhood from maternal and other close-contact sources. The principal cariogenic microorganisms are highly acidogenic and acid-tolerant bacteria and include members of the mutans streptococci group and lactobacilli. Numerous assessment tools for oral hygiene and caries risk are available to identify patients at risk for dental caries, but better quantitative methods are needed to allow appropriate targeting of aggressive, caries-protective treatments. Rapid adenosine triphosphate-driven (ATP-driven) bioluminescence assays have long been used to quantitatively measure microbial numbers in a variety of situations, including clean room monitoring and in the sanitation and food-processing industries. More recently, ATP-driven bioluminescence has been used to quantitate bacterial load in both medical and dental settings. For example, dental plaque mass and bacterial numbers in both humans and animals correlate well with ATP-driven bioluminescence values. The purpose of this review is to illustrate the use of ATP-driven bioluminescence in the rapid chair-side enumeration of total oral bacteria, including cariogenic plaque streptococci. Our studies using oral clinical specimens from pediatric patients demonstrate that ATP-driven bioluminescence can be used in the direct determination of bacterial numbers, and can serve as a general assessment indicator for

---

[*] Equal contributors to this work
[†] Corresponding Author: Curtis A. Machida, PhD. Department of Integrative Biosciences, Oregon Health & Science University, School of Dentistry; Telephone: 503-494-0034; Fax: 503-494-8554; machidac@ohsu.edu

oral hygiene. ATP-driven bioluminescence may potentially serve as a component of dental caries risk assessment.

Keywords: ATP-driven bioluminescence, dental and medical applications, cariogenic microorganisms and mutans streptococci, oral health, dental caries, caries risk assessment.

I. Introduction

A. ATP-Driven Bioluminescence and Microbiological Detection

Bioluminescence is generated by the conversion of chemical energy into light and is exhibited in diverse organisms, including select bacteria, cnidarians, dino-flagellates, fungi, fish, and insects [1]. The first observations of bacterial bioluminescence was made in the 17th century by Robert Boyle, who demonstrated that oxygen was required for light emission from "shining flesh" [2]. Later discoveries demonstrated that this light actually originated from bacteria. More recently, a variety of practical applications for bioluminescence have been developed, including its use for detection of microorganisms.

Almost all bioluminescent reactions involve the reaction of oxygen with luciferin substrates and luciferase enzymes that result in the production of visible light. The cloning of luciferase genes (*lux*) from prokaryotes permitted the genetic modification of non-luminescent bacteria to generate bioluminescence [3]. Once these organisms were bioluminescent, they could be easily detected by the presence of visible light. By using bioluminescence techniques, bacterial numbers and susceptibility to antibiotics could be rapidly determined [2,3]. This technology has allowed the development of assays for rapid detection of bacteria and determination of microbiological load [2].

Another type of bioluminescence that has considerable use in medicine and industry is adenosine triphosphate-driven (ATP-driven) bioluminescence. The ATP-driven bioluminescence assay is one of the most widespread methods in microbiology for rapidly detecting microorganisms [4]. By isolating intracellular ATP, living organisms, including bacteria, can be rapidly quantified [4]. Adenylated energy charge (AEC), the composite of ATP, ADP, and AMP, also correlates with bacterial number, is independent of growth phase, and can be measured with the use of the ATP-driven bioluminescence assay [5].

Like the bioluminescence emitted from genetically-modified microorganisms, the ATP-driven bioluminescence assay takes advantage of luciferase-luciferin enzyme-substrate reactions. The amount of ATP in microorganisms can be quantified by the release of light according to the following formula:

$$ATP + Luciferin + O_2 + Luciferase + Mg^{2+} \rightarrow AMP + oxyluciferin + PP_i + CO_2 + light$$

Bioluminescence assays have been subject to many recent improvements. For example, recombinants containing the luciferase gene from *Photinus pyralis* are capable of luminescence intensity ten-fold greater than the amount of light generated from wild type organisms [6]. In addition, background luminescence can be reduced through the use of adenosine phosphate deaminase, with AMP and pyrophosphate produced by the luciferase reaction being recycled to ATP using pyruvate orthophosphate dikinase (PPDK) [4]. This reduction in background luminescence results in dramatically increased amplification in observed luminescence [4], and allows the detection of a single bacterial cell containing approximately 10^{-18} mol of ATP [6]. Background ATP can also be reduced by isolation of intracellular ATP using filtration methods [3,4,7-15], and digestion of free ATP with ATPase enzymes [15]. Recently, detection and identification of bacteria has been enabled using a combined ATP-driven bioluminescence immunoassay [16].

In addition to many applications in medicine, bioluminescence, specifically ATP-driven bioluminescence, has been applied to the detection of microorganisms on hard surfaces [8,15], graphic documents [13], pharmaceuticals [11,17], water samples [9,12], and food products [3,7,10,14,16,18-20]. Its use for monitoring surface contamination has proven especially useful in the monitoring of clean rooms due to its ability to detect non-cultivable microbes [15]. The versatility of this assay is also demonstrated in its ability to distinguish between living and dead fungal strains on paper documents [13].

The adoption of ATP-driven bioluminescence in the pharmaceutical industry has been relatively slow [17]; however, this method is now sensitive enough to be used for preservative efficacy tests (PETs) [11]. PETs are used to detect bacteria, yeast, and mold in a single growth media, an advantage in this industry [11]. ATP-driven bioluminescence can also be used to detect microorganisms in water samples, and has been used to specifically test for contamination of *Escherichia coli* in beach water [12]. A similar procedure is also commonly used to test for both total bacterial and coliform bacteria contamination in drinking water.

ATP-driven bioluminescence has been used since the 1970s in the detection of microorganisms in food [19]. It has been used for bacterial detection on meat [8], poultry [8], beer [14], and fruits [8]. One of the benefits of the ATP-driven bioluminescence assay is its ability to deliver rapid results, making it well suited for spot monitoring [10]. This is illustrated by its use in the dairy industry. As in other industries, the dairy industry uses ATP-driven bioluminescence to monitor surface cleanliness at critical control points [3] including the assessment of contaminants on post-pasteurization equipment [20]. Gram-negative psychrotrophic bacteria are the primary causes of decreased shelf life of milk and are the primary sources of most contamination post-pasteurization [20]. Bioluminescence tests specific for the detection of bacterial ATP are also used to evaluate the preparation of milk and hygiene of cows for milking [18], thus ensuring the highest quality milk product.

B. Advantages and Limitations in the Use of ATP-Driven Bioluminescence in Microbiological Detection

1. Simplicity

ATP-driven bioluminescence assays are advantageous largely due to their simplicity. The ability to perform direct measurements without enrichment of samples dramatically shortens assay time and eliminates the need for bacterial culturing equipment. Automated assay readings based on light intensity reduce human error associated with traditional colony counting, and small sample sizes expedite collection and avoid dispersion of bacteria, therefore decreasing potential spread of disease [21]. Significant increases in efficiency and the bypassing of incubation requirements for cultures (both time and equipment) are considered the greatest advantages of the ATP-driven bioluminescence assay [22].

2. Accuracy

Most studies report that luciferin-luciferase reactions allow very precise determinations of ATP concentrations and bacterial load [23]. For example, ATP concentrations within bacteria have been determined by comparisons between ATP-driven bioluminescence and traditional plating methods [22]. However, some studies report ATP-driven bioluminescence readings exceed comparative conventional bacterial counts [24], while others report the presence of bacteria, confirmed by traditional methods, on surfaces considered "clean" by ATP-driven bioluminescence assessment [25]. Many of these inconsistencies appear to be based on differences between ATP levels in exponential and stationary growth stages [26], or presence of contaminating eukaryotic (or somatic) ATP [10]. Some studies have succeeded in increasing the sensitivity of the ATP-driven bioluminescence assays with the use of filtration and purification, although these methods tend to decrease the efficiency of the assay.

3. Efficiency

Efficiency is a significant advantage of the ATP-driven bioluminescence assay [23]. Traditional microbiological detection requires microbial incubation, enrichment, and isolation. While colony growth is still the "gold standard" of microbial detection, the extensive time and labor requirements places constraints on its use. The expedience and accuracy of the ATP-driven bioluminescence assay has made it a popular alternative to conventional microbial detection using colony formation.

The efficiencies of the ATP-driven bioluminescence assay also allow medical personnel to overcome the difficulties associated with slow growing fastidious microbes. Tuberculosis is a very common disease caused by the often antibiotic-resistant and slow growing bacterium *Mycobacterium tuberculosis*. The *in vitro* doubling time for *M. tuberculosis* is approximately 20 hours, making the determination of antibiotic resistance with the use of selective agars, slow and tedious. The use of ATP-driven bioluminescence circumvents the slow growth of *M. tuberculosis*, allowing for quicker determination of bacterial resistance and immediate development of an appropriate antibiotic regimen [27]. In addition, sterility of medical equipment is of pivotal importance for many medical procedures. The advent of the ATP-driven bioluminescence assay provides a rapid assessment of the sterility of medical equipment and operatories [28], therefore decreasing unnecessary microbial exposure to patients.

4. Specificity

While ATP-driven bioluminescence can be considered both quantitatively accurate and efficient, it is not able to differentiate the bacterial species being sampled. This non-discriminate characteristic is suitable for sanitation monitoring in which the determination of any microbial presence is desired; however, certain applications aimed at monitoring specific bacteria require the addition of isolation and purification protocols. Eukaryotic ATP from the host is also a problem for ATP assays. Since both prokaryotic and eukaryotic ATP produce the same bioluminescent activity using luciferin substrates, additional steps must be taken to eliminate eukaryotic ATP. Even though the elimination of eukaryotic ATP and isolation of specific bacteria are required for some assays, these measures require more time and may decrease sensitivity of the assay [10].

Nevertheless, substantial increases in efficiency, accuracy, and simplicity have made ATP-driven bioluminescence a keystone of modern microbiological detection. Rapid microbiological monitoring decreases potential pathogenic contamination, and is less expensive and demands a lesser time commitment than conventional microbiological detection alternatives.

C. Applications of ATP-Driven Bioluminescence in Medicine

Bioluminescence applications play significant roles in medicine, with new uses being continually designed. For example, oncology has recently begun to utilize bioluminescence. Cancer is characterized by uncontrolled growth of groups of cells, invasion into adjacent tissues, and potential metastasis to other regions of the body [29-34]. Cancer is often treated with chemotherapy using cytotoxic chemical agents. These drugs are designed to kill rapidly growing cells, often by hindering DNA replication [35]. Traditional chemosensitivity assays, including the thymidine uptake assay, measure radioactive DNA and RNA precursor incorporation. The thymidine uptake assay requires cells to be in the S phase of the division cycle, and cancer cells are not always in S phase, which limits the utility of these traditional chemosensitivity assays.

ATP-driven bioluminescence is an alternative method for measuring the effectiveness of cytotoxic agents *in vitro* [36]. ATP-driven bioluminescence has excellent correlation with other chemosensitivity assays and has the advantage of being rapid and quantitative [36,37]. This method involves culturing neoplastic specimens *in vitro*, followed by exposure of cultures with cytostatic drugs. ATP is then extracted and measured through the use of ATP-driven bioluminescence to evaluate drug effects. The quantity of ATP present indicates whether the cytostatic agent was effective in stopping or inhibiting neoplastic cell growth [36].

More recently, bioluminescence has been employed in tumor imaging with a technique called bioluminescence imaging (BLI), which can be performed both *in vivo* [38] and *in vitro* [39]. Light at low levels can be transmitted through tissues and detected with sensitive photon detection methods [40,41]. This enables the detection of optical signatures that are emitted by tumor cells by expression of reporter genes encoding luciferases. This method can be used to detect and understand molecular mechanisms of neoplasms [38], and has tremendous use in the assessment of cancer treatment response *in vivo* [42-44].

The luciferase gene from the *Photinus pyralis* firefly has been used in monitoring tumor growth and evaluating effectiveness of antineoplastic therapy in animal models of human neoplasia [45,46]. Luciferases from multiple sources, including marine mammals and marine organisms, have also been used. These luciferases cause the emission of blue (460 nm) light, and are used in dual reporter assays [47,48].

Traditional optical imaging often involves an external source of light to either examine the concentration of exogenous dyes that accumulate or are activated in tumor sites, or to investigate the optical properties of tumor tissue. In contrast, external detection of an internal bioluminescent signal offers several advantages [49-52]. Light originating from luciferase-labeled cells can reveal information with little or no background luminescence [53], whereas the background is much larger in the use of external light sources.

Low light imaging systems, such as charge-couple device cameras, can detect bioluminescent signals [41]. Reporter gene recombinants, upon transfer into tumor cells, are copied during cell division, and its use is advantageous over other techniques that involve the application of dyes that become diluted as cells divide. Using reporter genes, real time *in vivo* analyses can be achieved measuring pharmacological activity, pathogenic events, and/or promoter function [40,41,53]. Thus, reporter genes can allow for the viewing of biological processes in living animals, which can be used in the determination of the spatial and temporal distribution of neoplastic growth and metastasis *in vivo*. For example, the reporter cell line HeLa-*luc* was generated and introduced into severe combined immunodeficient (SCID) mice via intravenous, subcutaneous, or intraperitoneal inoculation, and the kinetics of proliferation of labeled cells were non-invasively monitored [54]. The signals were then used to quantify and rapidly locate tumors, with as few as 1000 labeled cancer cells quantitatively detected [54].

Bioluminescence as a reporter in living mammalian cells has also been utilized to evaluate gene expression [55-57], interactions between mammalian and bacterial cells [40,58], and viral infections [59-61], and has also been used in preimplantation mice embryos [55] and prokaryotic cells [62]. In other reports, our laboratory has recently developed and used bacterial *lux* recombinants in the transformation of oral streptococci, and has determined the potential antibacterial activity of minocycline hydrochloride in conjunction with anesthetics or the antiseptic chlorhexidine [63]. As discussed above, BLI has emerged as an effective method to analyze infectious diseases in animal models. Many studies of infection by bioluminescent Gram-positive and Gram-negative bacteria have been performed, and viruses and eukaryotic parasites, including *Leishmania, Plasmodium,* and *Toxoplasma*, have been genetically modified with luciferase genes to yield insight in their *in vivo* properties [64,65].

D. Dental Caries and Use of ATP-Driven Bioluminescence in Dentistry

Dental caries is one of the most prevalent diseases worldwide, with 42% of children and adolescents and 90% of adults in the United States having caries in permanent teeth [66]. Caries is a microbial disease that interacts in complex ways with individual factors, such as biology and diet. These factors affect the microorganisms' ability to cause disease and affect tooth structure [5,67]. The principal caries-causing microorganisms are mutans streptococci, including *Streptococcus mutans*, and lactobacilli both of which are normal components of oral microflora [68-70]. These microorganisms are able to initiate dental caries when large numbers adhere to the tooth surface, forming the biofilm that makes up dental plaque [71]. As the dental biofilm accumulates, the tooth surface becomes affected by the conversion of simple carbohydrates to weak organic acids and subsequent demineralization of the hydroxyapatite structure. The oral cavity has natural defense mechanisms in the buffering ability of saliva to neutralize these organic acids and an ability to stop and reverse demineralization. However, permanent destruction of tooth structure can occur when demineralization is occurring too frequently [69]. This destruction first becomes demonstrated as white spot lesions that can progress to the formation of cavitated lesions [70].

In dentistry emphasis is placed not only on removing and filling the damaged tooth structure, but also on predicting and preventing the disease. Assessment tools have been developed and refined that address multiple components involved in the progression of caries [68]. Methods for assessment of the presence of microorganisms in the oral cavity are available, including the measurement of lactic acid on the tongue, reflective of cariogenic bacterial cell number [72], and quantitative assessment of cariogenic streptococci by competitive polymerase chain reaction [73,74]. In addition, several bioluminescence assays have been designed to quantitate plaque mass in human and animal specimens, including assays that measure ATP, flavin mononucleotides, phosphocreatine, and adenylated energy charge [21,24,75,76]. While some assays focus on the quantification of *S. mutans* and lactobacilli in saliva [77-79], other assays target applications in dental plaque [80]. ATP-driven bioluminescence is emerging as a tool in the dental office by providing quantitative measurement of microorganisms that are associated with dental caries.

There are several benefits for using ATP-driven bioluminescence in the dental field. First, as defined above, ATP-driven bioluminescence measurements could be used in the chair-side quantification of bacteria contained within dental plaque, and thus can determine the component risk factor for dental caries, and can also be used to evaluate and track the oral hygiene of a patient. By being able to quantify bacteria in dental plaque, this number could be divulged to the patient, providing positive reinforcement when conditions are improving and assisting in the illustration of potential problems. Any variation in bacterial numbers could indicate a change in the oral environment that may lead to the potential initiation of dental caries [67]. Second, ATP-driven bioluminescence measurements can also help clinicians evaluate effectiveness of an anti-microbial regimen, product and treatment plan [76]. Lastly, because the ATP-driven bioluminescence method is extremely sensitive, it can measure the viable cell mass from a single tooth [21]. The sensitivity of this method may be important in monitoring specific problem areas of the patients' dentition, and tracking levels of plaque bacteria at specific sites [21,67]. In total, ATP-driven bioluminescence as a tool for the quantification of bacteria present in dental plaque can potentially be used in the assessment of caries risk, patient treatment and home care, and as an auxiliary aid for patient education.

E. Biochemical and Microbiological Determinants of Dental Caries

Dental caries is a disease whereby bacteria cause damage to hard tooth structure, including enamel, dentin, and cementum. Teeth are comprised of an outer mineralized epidermal tissue or enamel. The organic matrix of the enamel is secreted from ameloblast cells in the enamel organ of the developing tooth. Calcium ions form seeds of inorganic hydroxyapatite [HAP; $Ca_{10}(PO_4)_6(OH)_2$] throughout the developing matrix, which grow and supplant the organic matrix, until the matrix is comprised of 96% inorganic matrix consisting of repeating crystalline structure of enamel. However, most of the apatite is in an impure form, containing substantial amounts of carbonate in the lattice. Exposure to acids demonstrates that carbonated components are most susceptible to demineralization and the first to be dissolved. Fluoride substitution in the apatite lattice provides a therapeutic effect of creating fluoroapatite or fluorohydroxyapatite, which is significantly less soluble than HAP. Fluoride also facilitates remineralization following the cycle of acid dissolution.

Underlying the enamel is a tissue called dentin, which is mainly composed of thin apatite crystal flakes creating dentinal tubules encased within a protein matrix of cross-linked collagen fibrils. Odontoblast cells have extended processes within each of the dentinal tubules, with its cell body present in the dental pulp, which is the site of most of the neurovascular tissues of a tooth. Dentin is comprised of 70% inorganic matrix (HAP), 20% organic (mainly Type I collagen), and 10% water by weight. If dental caries extends into the dentin, this is often accompanied by the feeling of dental pain, as the dentinal tubules contain water which can conduct noxious substances, with pressure and temperature changes to nerve fibers present in the central pulp. Dentinal damage and collapse is also the cause of cavitation of the external enamel surface that often accompanies dental caries. Caries often occurs in the cementum, which is a bone-like tissue that is avascular and covers the surfaces of dental roots in the place of enamel.

The initial attachment of bacteria to the host tissue is the first step for infection and colonization. The attachment step is facilitated by interactions with cell-surface adhesin-like molecules [81,82]. *S. mutans* contains adhesin-like molecules that allow binding to the negatively-charged acquired enamel pellicle (an acidic glycoprotein layer) of teeth, and with the presence of sucrose, subsequently produce dextran polysaccharides that are strongly adhesive [81,83]. MS have the ability to consume common table sugar (sucrose) and convert sucrose via any one of three isozymes of glucosyltransferase (GTFs) into dextran (n-glucose). MS-derived dextran is a complex glucan that is known for its strong adhesiveness. This dextran material holds cells together as they proliferate and divide, forming very sticky plaque. *S. mutans* has also been demonstrated to produce three glucan-binding proteins, which also mediate adherence [83]. Carious lesions occur primarily by dissolution of mineral in enamel and dentin, due to acids released by these bacteria [84]. As a facultative anaerobe, *S. mutans* can survive in the presence of oxygen and ferment glucose into lactic acid, which decreases local pH and causes enamel erosion [81]. Other virulence factors of MS include their ability to ferment sucrose into lactic acid more rapidly than other oral bacteria [85]. Some *S. mutans* strains also have the ability to promote both intracellular and extracellular polysaccharide storage, and can continue metabolism in the absence of external food supply [86]. The acid tolerance of certain *S. mutans* strains is another virulence factor, as the decreased pH due to lactic acid formation would otherwise diminish *S. mutans* populations [87]. Finally, *S. mutans* may have the ability to reduce the presence of competing bacteria by secreting bacteriocins, compounds which inhibit the growth of other bacterial species, thus aiding in the establishment of these cariogenic bacteria [82].

Lactobacilli can also be highly acidogenic and acid tolerant [88], but are known to be late colonizers of the mouth [89-91]. Some data suggest that lactobacilli are more likely to colonize near a pre-existing colonization of MS, and thus are not requisite for the initiation of lesions but rather contribute to its continuation. Thus, they may play an important role in continued demineralization once caries are established [92-94].

II. METHODS

A. Use of ATP-Driven Bioluminescence to Assess Dental Caries Risk in Pediatric Patients

One of the major issues in pediatric dentistry is developing rapid, reliable, chair-side assessment tools for dental caries risk. This study was conducted to determine if ATP-driven bioluminescence could be used in the measurement of total oral and plaque bacterial load, and more importantly cariogenic MS numbers, and as a potential assessment tool for dental caries risk in pediatric patients.

Table 1. Patient Demographics and Clinical Obaservations[1]

Gender		**Age**		**Ethnicity**		**Residence**		**Medical History[2]**		**Additional Fluoride Use[3]**	
	Number of Patients		Number of Patients		Number of Patients		Number of Patients		Number of Patients		Number of Patients
Male	23	7 yrs	9	African-American	3	City of Portland	12	ASA 1	29	In Water	7
Female	7	8 yrs	7	Asian	3	Portland Suburbs	14	ASA 2	1	Supplemented	5
Patient Status		9 yrs	4	Caucasian	15	Other	4				
New	22	10 yrs	3	Hispanic	9						
Recall	8	11 yrs	4								
		12 yrs	3								
Oral Hygiene[4]		**Gingivitis[5]**		**Number of Cavitated Teeth**		**Restorations**		**Toothbrushing Frequency**		**Hours After Brushing Before Visit**	
	Number of Patients		Number of Patients		Number of Patients		Number of Patients		Number of Patients		Number of Patients
Good	3	Mild	12	None	9	None	22	3x daily	2	0-1 hrs	8
Fair	13	Moderate	17	1 to 2	7	Amalgam	6	2x daily	21	2-3 hrs	20
Poor	14	Severe	1	3 to 5	7	Composite	1	1x daily	6	>3 hrs	2
				6 or more	7	Sealant	1	<1x daily	1		

[1] The OHSU Pediatric Dentistry Clinic generally serves patients of low socioeconomic status. Chart recordings listed medications, including fluoride tablets, oral rinses, and antibiotic use within the last 30 days, last tooth brushing, oral hygiene habits, patient status (new or recall patient), plaque index, and hygiene/tissue condition, or presence of gingivitis and periodontal disease.

[2] American Association of Anesthesiologists (ASA) physical status classifications: ASA 1 – healthy patient with no medical problems; ASA 2 – patient with mild systemic disease, in this case, one patient had mild controlled asthma.

[3] Additional fluoride use means supplemented fluoride in addition to fluoridated toothpaste. Thus, seven patients had access to fluoridated water and five patients had supplemented fluoride from sources other than fluoridated toothpaste. All 30 patients had access to fluoridated toothpaste.

[4] Definitions for oral hygiene: Good – pink gingiva, surfaces free of debris; Fair – red/pink gingiva, surfaces had some debris; Poor – red gingiva, surfaces had definite debris.

[5] Definitions for gingivitis: Mild – marginal gingivitis; Moderate – papillary gingivitis; Severe – spontaneous bleeding of gingiva and/or periodontal disease is evident.

B. Clinical Observations, Plaque Index Score and Caries Activity

Patients (33) who were registered for dental care in the Pediatric Dentistry Clinic of the Oregon Health & Science University (OHSU) were randomly selected for inclusion in this study. The inclusion criteria for this study were age (7-12 years old) and demonstrated good health. The 7-12 year old age group exhibited a wide range of caries experience and plaque levels. The criteria for exclusion in this study were individuals undergoing orthodontic treatment, because appliances can modify tooth surface characteristics, and individuals using any saliva or diet-altering medications. All protocols, including use of consent forms and specimen collection procedures, were reviewed and approved by the OHSU Institutional Review Board. Chart recordings listed medications, including fluoride tablets, oral rinses, and antibiotic use within the last 30 days, plaque index, hygiene/tissue condition, and presence of gingivitis and periodontal disease (Table 1). The oral examination included indication of missing teeth, decayed teeth and restored teeth.

Plaque index scores were completed on selected teeth for all patients. The presence of plaque was determined by visual examination and by tactile probing using an explorer instrument. Four surfaces for each selected tooth were examined, and the number of plaque-free surfaces within each tooth was assessed for each patient. Caries activity was also recorded based on clinical evaluation.

C. Plaque Specimens Collected from Specific Teeth

Four teeth representing all four quadrants of the mouth were tested in each patient. These four teeth represented both permanent and primary teeth with different surfaces, which were selected based on the age of the patient population, and on the probability of selected teeth being present at the time of examination. The upper right first molar and lower left molar were chosen because of their difficulty to brush for most patients, and also because of the close proximity to a salivary duct for the upper right first molar and the distance away from the duct for the lower left molar. An upper incisor was chosen because of its susceptibility to show enamel demineralization and significant plaque accumulation in children [95-97]. Finally, a lower incisor was chosen because of its close proximity to the salivary gland and access to the enhanced cleansing properties of the tongue.

Individual disposable picks (Opalpix; UltraDent Products, Inc.) were used to collect samples of dental plaque from the buccal surface of the upper first molar (#3 Buccal), the facial surface of the upper left anterior tooth (#9 Facial or #H Facial), the lingual surface of the lower left premolar or second primary molar (#20 Lingual or #K- Lingual), and the lingual surface of the lower central incisors (#25 Lingual). The collection sites were chosen from first preference site to next available preferred site. Collection of plaque from secondary preferred sites occurred in nine out of the 30 participants because of unerupted or early loss of the first tooth site.

The collection was conducted by a sweeping action across the entire tooth surface. After scoring of plaque index and plaque collection, the participants were given paraffin wax tablets for chewing and collection of stimulated saliva.

D. Microbiological Identification of Plaque and Saliva Bacteria

Plaque and saliva specimens were plated on enriched blood agar (PML Microbiologicals, Wilsonville, OR) to determine total bacterial numbers. Total oral streptococci were determined by limiting dilution plating on mitis salivarius agar (MSA, Difco^{TM}; Becton, Dickinson and Company, Sparks, MD), containing high sucrose and vital dyes (i.e.: crystal violet and bromphenol blue) and supplemented with potassium tellurite (Difco^{TM}). In order to select and enumerate the sub-group of mutans streptococci (MS), MSA including potassium tellurite, was further supplemented with bacitracin (10 Units / ml; Sigma Chemical, St. Louis, MO). All platings were conducted in quadruplicate, and plates exhibiting colony numbers between 50-500 were counted and averaged to determine mean values. As an additional validation in the determination of cariogenic bacteria, we used the commercial CRT Bacteria Test (Vivadent, Lichtenstein, Germany), which incorporates MSA and Rogosa agar in a dual slide test. CRT readouts to determine MS or lactobacilli numbers in saliva were interpreted as either $\leq$ or $\geq 10^5$ bacteria / ml.

E. ATP-Driven Bioluminescence Determinations from Plaque and Saliva Specimens

ATP-driven bioluminescence measurements from bacteria contained in plaque or saliva specimens were determined with the BacTiter Glo Microbial Cell Viability Assay (Promega, Madison, WI). ATP-driven bioluminescence was measured using the Veritas microplate luminometer (Turner Biosystems). BacTiter Glo contains thermostable luciferases and a proprietary formulation for extracting ATP from bacteria, and generates a "glow-type" luminescence signal from the luciferase reaction with a half-life of $>$ 30 minutes. The Veritas luminometer has a 10^5-fold dynamic range in relative light unit (RLU) readouts. In parallel luminescence assays, we utilized the CariScreen ATP bioluminescence swab collection system and hand-held luminometer (Oral Biotech, Albany, OR); this collection device consisted of a swab and swab holder for collection of oral specimens, and an upper reservoir containing luciferin, luciferase, and extraction components which can be drained over the swab following collection of specimens. The luciferase contained in the CariScreen system is based on the "flash-type" luminescence signal with RLU readouts peaking at two minutes and diminishing thereafter.

II. Results

A. Calibration of ATP Standards to Bioluminescence Measurements

Using ATP standards in picomolar (pM) or nanomolar (nM) concentrations, bioluminescence (RLUs) standard curves were developed for both the Veritas luminometer and CariScreen ATP Meter (Figure 1A, left and right panels, respectively). The Veritas luminometer has $> 10^5$ dynamic range measuring ATP from 10 pM to 1 μM (with RLU readouts from 10^4 - 10^8 RLUs), and the CariScreen meter has approximately 100-fold dynamic range measuring ATP from 100 pM – 10 nM (with RLU readouts from 15 – 4000 RLUs).

B. Statistical Correlations Linking ATP-Driven Bioluminescence Values to Total Oral Bacteria, Total Oral Streptococci and Total Mutans Streptococci Numbers

Saliva and plaque specimens were collected from selected teeth of 33 patients (n=4 plaque specimens per patient), and ATP-driven bioluminescence was measured from each specimen. Using serial dilution plating of each oral specimen, quantitation was also conducted for total bacteria on enriched blood agar plates and total streptococci and mutans streptococci on selective medium (MSA), in the absence or presence of bacitracin, respectively. Using the Veritas luminometer, ATP-driven bioluminescence values were determined using both plaque and saliva specimens. Significant Pearson correlation coefficients of 0.854, 0.840, and 0.796 were determined for total oral bacteria, total oral streptococci, and mutans streptococci, respectively (see Figure 1B, left, middle and right panels, respectively). When these ATP-driven bioluminescence readings were analyzed using plaque specimens only, which reduces the statistical power because of a smaller sample set, significant Pearson correlation coefficients of 0.682, 0.611, and 0.548 were still identified for total oral bacteria, total oral streptococci, and mutans streptococci, respectively (Figure 1C, left, middle and right panels, respectively). When scatter plot analyses were conducted correlating total oral bacteria with either total oral streptococci or MS (Figure 1D; left and right panel, respectively), increasing numbers of total oral bacteria were found to track linearly with total oral streptococci in a strongly significant relationship ($r = 0.942$), and also to a lesser but still significant degree with MS ($r = 0.700$).

Similar ATP bioluminescence relationships were found using the handheld CariScreen ATP Meter, where bioluminescence readouts for composite plaque and saliva specimens, or for plaque specimens alone, correlated well with numbers determined for total oral bacteria, total oral streptococci, and mutans streptococci (see Figure 2A and 2B; r values of 0.810, 0.780 and 0.753, respectively, using plaque and saliva specimens, and r values of 0.587, 0.509, and 0.471, respectively, using plaque specimens only).

Thus, ATP-driven bioluminescence is highly predictive of the numbers of total oral bacteria, total oral streptococci and total MS. Strong statistical correlations were determined using either the Veritas luminometer or the CariScreen ATP Meter, but only when using the increased statistical power of the larger sample number contained in the composite plaque and saliva specimen set.

CRT scores for both MS and lactobacilli numbers were also recorded for each saliva specimen. While CRT scores are based on a scale of 1-4 (low to high: scores 1 and 2 being $< 10^5$ colonies and 3 and 4 being $> 10^5$ colonies) and are considered to be semi-quantitative evaluators of MS content, the CRT scores obtained for the saliva specimens were consistent with bacterial cell numbers enumerated by our direct plating on selective agars (Figure 3A).

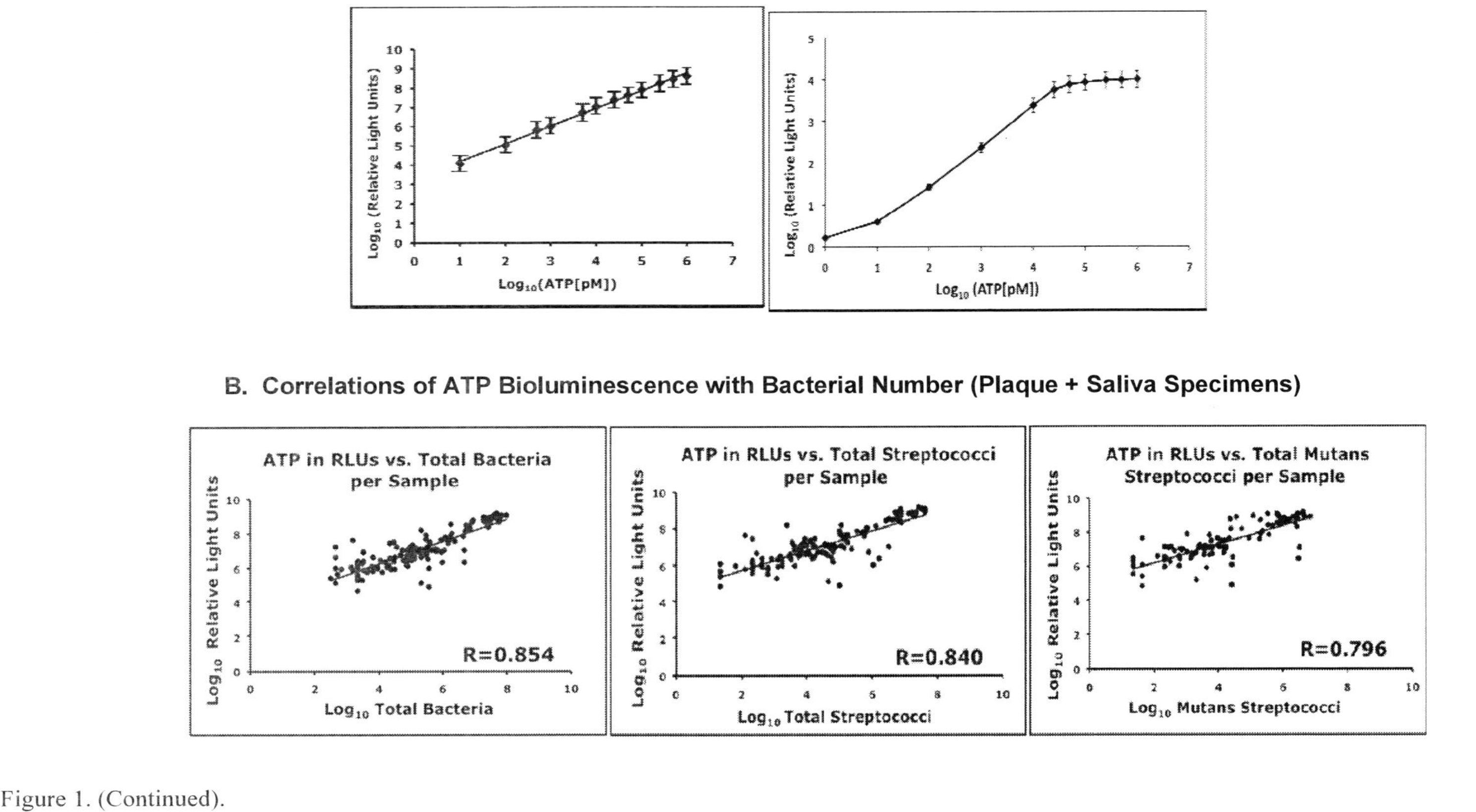

Figure 1. (Continued).

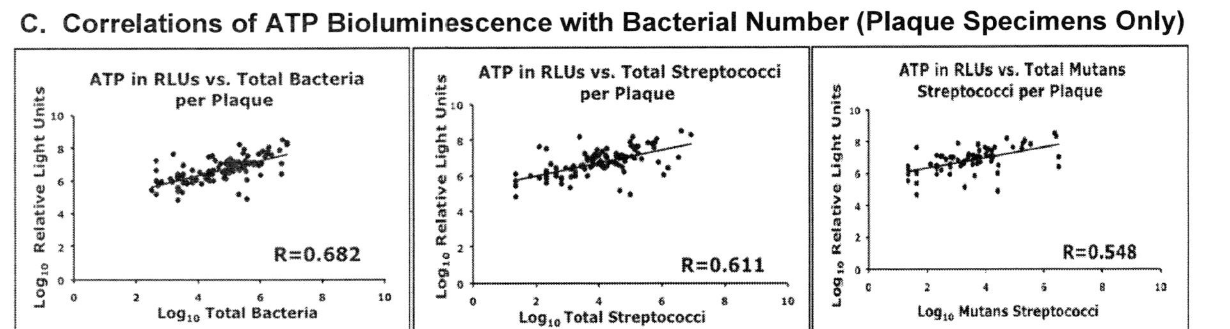

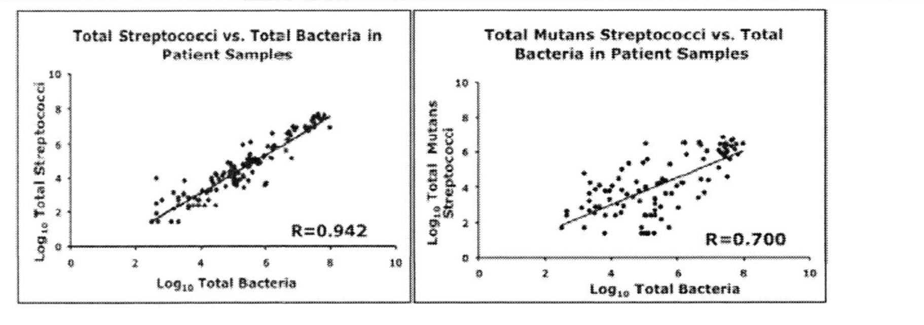

Figure 1. Panel A. ATP standard curves comparing ATP concentration to bioluminescence readouts (RLUs) using the Veritas microplate luminometer (left panel; error bars represent 1 standard error; n = 5 replicate determinations) and CariScreen ATP Meter (right panel). Panels B and C. Scatter plot analysis correlating ATP-driven bioluminescence (derived from the Veritas microplate luminometer) versus bacterial cell number for total oral bacteria (left panels), total oral streptococci (middle panels), and MS (right panels). Panel B depicts data for collection of plaque and saliva specimens and Panel C depicts data for plaque specimens only. Both Panels B and C contain data from specimens collected from 30 participants examined in 2007. N values equals 30 patients x 4 plaque specimens = 120 plaque specimens plus 30 additional saliva specimens. Panel D depicts scatter plot analysis correlating total oral bacterial numbers to total oral streptococci (left panel) and MS (right panel). Data points containing measurements of bacterial number in all plots are the mean values of 4 replicates (n=4) using plating dilutions exhibiting 50-500 colonies. ATP measurements are tabulated as the mean of 4-5 replicate and parallel determinations conducted with the Veritas luminometer. Pearson correlation coefficients (*r* values) are noted in each panel for all correlations. This figure and figure legend were reproduced with the kind permission of the AAPD from our article [Fazilat *et al.* (2010) Application of adenosine triphosphate-driven bioluminescence for quantification of plaque bacteria and assessment of oral hygiene in children. *Pediatric Dentistry* 32(3):195-204].

A. Correlations of ATP Bioluminescence (CariScreen) with Bacterial Number (Plaque and Saliva)

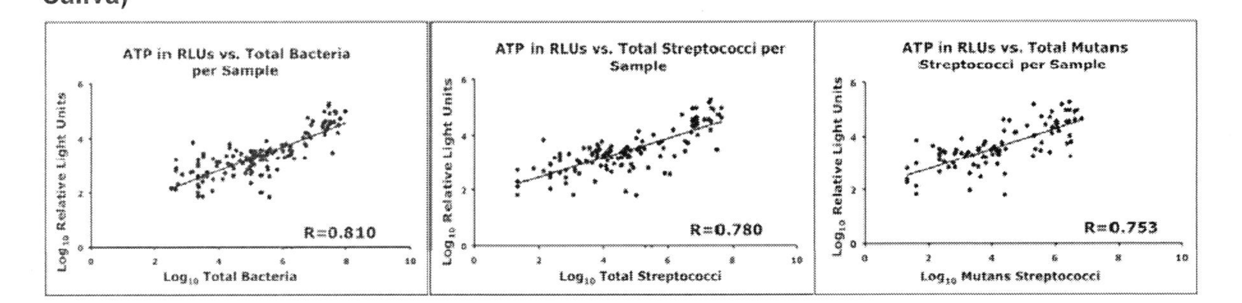

B. Correlations of ATP Bioluminescence (CariScreen) with Bacterial Number (Plaque Only)

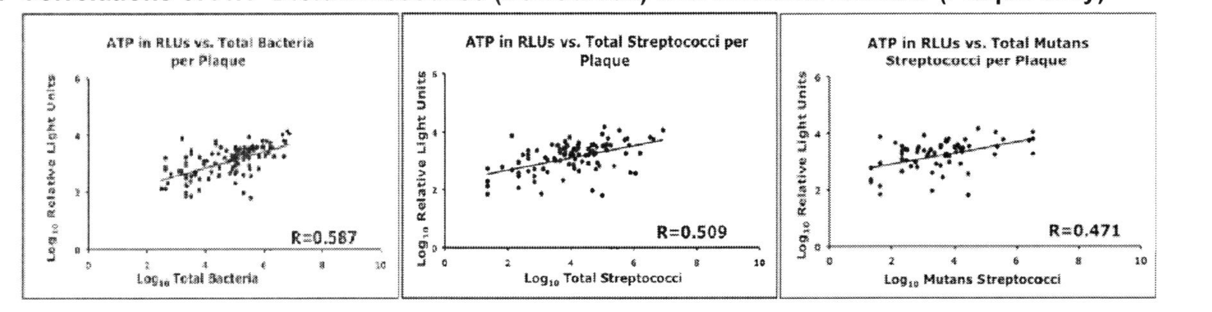

Figure 2. Panels A and B. Scatter plot analysis correlating ATP-driven bioluminescence (derived from the CariScreen ATP Meter) versus bacterial cell number for total oral bacteria (left panels), total oral streptococci (middle panels), and MS (right panels). Panel A depicts data for collection of plaque and saliva specimens and Panel B depicts data for plaque specimens only. This figure displays data from specimens collected from 30 participants examined in 2007. Typical n values equals 30 patients x 4 plaque specimens = 120 plaque specimens plus 30 saliva specimens. Data points containing measurements of bacterial number in all plots are the mean values of 4 replicates (n=4) using plating dilutions exhibiting 50-500 colonies. ATP-driven bioluminescence measurements are tabulated as the mean of 4-5 replicate and consecutive determinations conducted with the CariScreen ATP Meter. Pearson correlation coefficients (*r* values) are noted in each panel for all correlations. This figure and figure legend were reproduced with the kind permission of the AAPD from our article [Fazilat *et al.* (2010) Application of adenosine triphosphate-driven bioluminescence for quantification of plaque bacteria and assessment of oral hygiene in children. *Pediatric Dentistry 32*(3):195-204].

A.

Lactobacilli CRT Score[1]		Mutans CRT Score[1]		Plaque – Free Surfaces[2]		Active Decay Surfaces[3]	
Score	Number of Patients	Score	Number of Patients	Score	Number of Patients	Score	Number of Patients
1	5	1	4	>80%	0	None	9
2	7	2	3	50-79%	15	1–4	7
3	10	3	10	21-49%	11	5-10	7
4	5	4	10	>20%	4	>11	7

[1] CRT scores were enumerated from only 27 patients because of temporary unavailability of CRT kits from manufacturer and domestic suppliers.

[2] Plaque index scores on selected teeth were determined for each patient. The use of visual examination (without the use of disclosing solution) and tactile feel by an explorer instrument was utilized to detect plaque presence. The number of plaque-free surfaces (out of four possible plaque-free surfaces) were counted for each tooth, and the percentage of plaque-free surfaces were calculated for each patient. After plaque collection and scoring of plaque index, the participant was instructed to chew a paraffin wax tablet and expel saliva into a sterile collection container.

[3] Active decay surfaces were determined by visual examination alone, and not based on use of radiographs.

Figure 3. (Continued).

B.

OHSU Caries Risk Index

1. Lactobacilli CRT Numbers Score

Cell Number	$<10^4$	$>10^4<10^5$	$>10^5<10^6$	$>10^6$
Score	1	2	3	4

2. S. mutans CRT Numbers Score

Cell Number	$<10^4$	$\geq 10^4<10^5$	$\geq 10^5<10^6$	$>10^6$
Score	1	2	3	4

3. Plaque Index Score

% Plaque-free Surfaces	>80%	50-79%	21-49%	<20%
Score	1	2	3	4

4. Active Surface Decay Score

# Decayed Surfaces	None	1-4	5-10	≥11
Score	1	2	3	4

Final Caries Risk Level

Sum of Scores	4-6	7-9	10-16
Risk Level	Low	Medium	High

C.

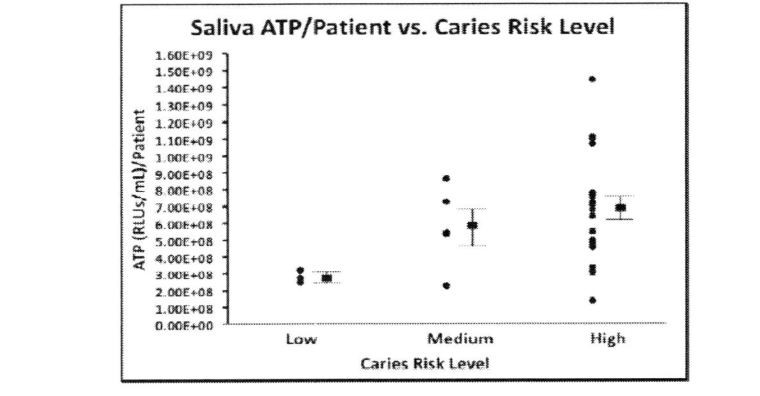

Figure 3. Panel A. Clinical observations, including CRT scores and plaque-free and active decay surfaces, used for determination of OHSU Pediatric Dentistry Caries Risk Index. Panel B. Scoring chart for OHSU Pediatric Dentistry Caries Risk Index. Panel C. Saliva ATP/patient values versus caries risk level. Caries risk level is a composite score (low risk: range 4-6; medium risk: 7-9 and high risk: 10-16) of traditional plaque index used by dental professionals, number of active decay (cavitated) surfaces, and CRT results. Individual plaque scores of 1-4 are based on the percentage of plaque-free surfaces. The active decay scores of 1-4 are based on the number of cavitated surfaces. The CRT result, divided as scores for mutans streptococci and lactobacilli, is a graded ranking of bacterial numbers of MS and lactobacilli. Values from all categories are then added to develop the composite OHSU Pediatric Dentistry caries risk index score, and then designated as low, moderate, or high caries risk. Panel C displays data from 33 participants, including 30 participants seen in 2007 and 3 additional low caries risk patients seen in 2008. n=5 for low caries risk individuals. This figure and figure legend were reproduced with the kind permission of the AAPD from our article [Fazilat *et al*. (2010) Application of adenosine triphosphate-driven bioluminescence for quantification of plaque bacteria and assessment of oral hygiene in children. *Pediatric Dentistry 32*(3):195-204].

III. Discussion

A. Criteria Used for Development of OHSU Pediatric Dentistry Caries Risk Index

The majority of patients seen at the OHSU Pediatric Dental Clinic would be considered high risk according to the American Academy of Pediatric Dentistry (AAPD) guidelines. Furthermore, the AAPD holds previous caries-related history as a bias against potential reassessment of future caries risk. In order to better rank our study population, we composed quantifiable clinical and microbiological factors for determination of current caries risk.

Three index factors were used to develop the OHSU Pediatric Dentistry caries risk index (CRI). The first index factor was visible bacterial plaque. Visible plaque on labial surfaces of maxillary incisors has been found to be strongly associated with caries development [98]. Children with no visible plaque at two years of age had greater chances of remaining caries-free until three years of age, compared to children with visible plaque [99]. Also, professional plaque removal could prevent caries, thus establishing dental plaque as a significant and probable risk factor [100-102].

The second index factor was presence of active caries. Previous caries experience was an important predictor in most models tested for primary, permanent, and root surface caries [103]. For permanent teeth in children and adolescents, DMFS (decayed, missing, filled surfaces) scores and pit and fissure morphology were the most important indicators. Caries in the primary dentition was also predictive of caries in the permanent dentition [104]. Furthermore, 84% of the children who did not display caries in the primary dentition remained caries-free in the mixed dentition [105].

The third index factor was levels of MS and lactobacilli. MS is an important determinant in the development of dental caries [81,102,106,107]. Caries-active individuals, compared to caries-free individuals, harbor higher numbers of MS and lactobacilli in their saliva and plaque [102,106,108-113]. In longitudinal studies, the numbers of MS and lactobacilli increase with the onset and progression of caries [92,114-116]. Another important risk factor is age at time of significant colonization with MS. Infants with high levels of MS, compared to older children, have more severe caries in the primary dentition [98,117,118]. Most infants acquire significant MS between 19-33 months of age, with some as early as 8-10 months. As a result, the AAPD, American Dental Association (ADA), and the American Association of Public Health Dentistry (AAPHD) all recommend oral evaluation of children by one year of age. The American Academy of Pediatrics (AAP) recommends referral to a dentist by age 1 if the mother has high caries rate, the child has demonstrable caries, plaque, demineralization, and/or staining, the family is low socioeconomic status, the child has special care needs, or if the child is a later order offspring [119].

B. Quantitative Formulation of OHSU Pediatric Dentistry Caries Risk Index

The OHSU Pediatric Dentistry caries risk index score reflects measures consisting of 1) traditional plaque index used by dental professionals, 2) number of cavitated surfaces, and 3) CRT (Caries Risk Test, Vivadent) results quantifying numbers of MS and lactobacilli (Figure

3A and 3B). Values from all three categories were then totaled to develop the composite caries risk score, and then designated as low, medium, or high caries risk using the following equation:

OHSU Pediatric Dentistry Caries Risk Index = Plaque Score (1/4) + Active Surface Decay Score (1/4) + CRT (includes composite S. mutans and Lactobacilli scores) (1/2)

Thus, the ranges of composite scores for low caries risk, medium caries risk and high caries risk were 4-6, 7-9 and 10-16, respectively (Figure 3A and 3B). By combining the three risk factors, the goal was to develop a quantitative assessment tool using indicators that were more reflective of current patient status, for direct comparison to measurements of ATP-driven bioluminescence identified in this study.

C. Relationship between ATP-Driven Bioluminescence, Oral Hygiene and Caries Risk

There appears to be stepwise increase in ATP-driven bioluminescence (obtained from saliva specimens) for any given patient when plotted against the broad categories of patients exhibiting low and medium / high caries risk based on our OHSU Pediatric Dentistry caries risk index score (Figure 3C). Even though the majority of individuals seen at the OHSU Pediatric Dental Clinic were considered as high caries risk, the sample size for low caries risk individuals in this study was sufficient (n=5) for making comparative determinations. We intend to significantly increase the sample size, including more low risk individuals, in future studies. In addition, based on ATP-driven bioluminescence values alone, we were unable to segregate individuals between the medium and high caries risk categories, and consider patients to be either in the category of low caries risk or the broad category of medium / high caries risk. We were also unable to discriminate differences between primary and permanent teeth, and plan to explore these potential differences in future studies.

While we understand that our ATP-driven bioluminescence values measure total oral bacterial number and not necessarily cariogenic microorganisms, this assay represents an excellent assessment tool for the efficacy of oral hygiene and potentially for the use of interventional procedures, including the use of anti-bacterial mouth rinses [63]. Based on our statistical analysis, patients with high plaque bacterial load also contain comparatively high streptococci and cariogenic MS numbers. While not absolutely definitive for cariogenic streptococci, the use of ATP-driven bioluminescence may potentially be used for assessment of plaque bacterial number and plaque load as an indirect indicator of dental caries risk.

CONCLUSION

1. ATP-driven bioluminescence may be used as a quantitative assessment tool for the determination of total oral bacteria, including plaque bacteria, and can be used to assess the efficacy of oral hygiene.

2. The strong statistical correlation between total oral bacteria to oral streptococci, including cariogenic streptococci, suggests that ATP-driven bioluminescence may also potentially serve as an indirect assessment determinant of dental caries risk.
3. Because the majority of pediatric patients seen at the OHSU Pediatric Dentistry Clinic are characterized as high caries risk based on previous caries history, we developed a quantitative caries risk assessment tool for pediatric patients that combined clinical and microbiological evaluations to help better define current risk for dental caries.
4. ATP-driven bioluminescence has broad implications in dentistry and medicine, and can be used to determine the efficacy of interventional therapies, including the use of mouth rinses, and in the detection of periodontal infections and other infectious diseases.

Acknowledgments

IK, CWK, JP, AT and AHV were all equal contributors to this work. IK was a 2010 OSLER student research fellow, and AV was a 2010 American Association for Dental Research student research fellow. IK obtained support from the Oregon Clinical and Translational Research Institute (OCTRI), grant numbers TL1 RR024159 and UL1 RR024140 from the National Center for Research Resources (NCRR), a component of the National Institutes of Health (NIH), and NIH Roadmap for Medical Research. CWK, JP and AT obtained support as 2010 summer research scholars from Dean Jack Clinton, OHSU School of Dentistry. IK, CWK, JP, AT and AV are all dental students at the OHSU School of Dentistry. TM and CM are faculty supported by the OHSU School of Dentistry. The authors thank Drs. Tom Shearer, John Engle and Jack Clinton for their encouragement and support in the development of this research program. The authors thank Dr. Kim Kutsch and Oral Biotech for the luminometer used in selected aspects of this study, and the OCMID group for support of the dental students. The authors thank the American Association for Pediatric Dentistry (AAPD) for granting copyright release to re-use text from our manuscript published in *Pediatric Dentistry* [Fazilat *et al.*, (2010) Application of adenosine triphosphate-driven bioluminescence for quantification of plaque bacteria and assessment of oral hygiene in children. *Pediatric Dentistry* 32(3): 195-204], including material from the Methods, Results, Discussion and Conclusion sections, as well as Table 1 and Figures 1, 2, and 3. Substantial text from the *Pediatric Dentistry* manuscript was reproduced in this new work, and the table and figures were reproduced in entirety. We also acknowledge and thank the authors of the original work, including Drs. Shahram Fazilat, Rebecca Sauerwein, Jennifer McLeod, Tyler Finlayson, Dr. Emilia Adam, Dr. John Engle, and Dr. Prashant Gagneja, (also TM and CM), for allowing us to reproduce this work. The authors thank Drs. Elizabeth Palmer and John Peterson, faculty members of the Department of Pediatric Dentistry, OHSU School of Dentistry, for the review of this manuscript.

REFERENCES

[1] Wilson T, Hastings JW. Bioluminescence. *Annu Rev Cell Dev Biol* 1998;14:197-230.

[2] Ulitzur S. Established technologies and new approaches in applying luminous bacteria for analytical purposes. *J Biolumin Chemilumin* 1997;12:179-192.

[3] Griffiths MW. Applications of bioluminescence in the dairy industry. *J Dairy Sci* 1993;76:3118-3125.

[4] Sakakibara T, Murakami S, Imai K. Enumeration of bacterial cell numbers by amplified firefly bioluminescence without cultivation. *Anal Biochem* 2003;312:48-56.

[5] Fazilat S, Sauerwein R, McLeod J, Finlayson T, Adam E, Engle J, Gagneja P, Maier T, Machida CA. Application of adenosine triphosphate-driven bioluminescence for quantification of plaque bacteria and assessment of oral hygiene in children. *Pediatr Dent* 2010;32(3):195-204.

[6] Noda K, Matsuno T, Fujii H, Kogure T, Urata M, Asami Y, Kuroda A. Single bacterial cell detection using a mutant luciferase. *Biotechnol Lett* 2008;30:1051-1054.

[7] Brovko L, Froundjian V, Babunova V, Ugarova N. Quantitative assessment of bacterial contamination of raw milk using bioluminescence. *J Dairy Research* 1999;66:627-631.

[8] Chen FC, Godwin SL. Comparison of a rapid ATP bioluminescence assay and standard plate count methods for assessing microbial contamination of consumers' refrigerators. *J Food Prot* 2006;6(10):2534-2538.

[9] Frundzhyan V, Ugarova N. Bioluminescent assay of total bacterial contamination of drinking water. *Luminescence* 2007;22:241-244.

[10] Gracias KS, McKillip JL. A review of conventional detection and enumeration methods for pathogenic bacteria in food. *Can J Microbiol* 2004;50:883-890.

[11] Jimenez L. Molecular diagnosis of microbial contamination in cosmetic and pharmaceutical products: a review. *J AOAC Int* 2001;84(3):671-675.

[12] Lee J, Deininger R. Detection of *E. coli* in beach water within 1 hour using immunomagnetic separation and ATP bioluminescence. *Luminescence* 2004;19:31-36.

[13] Rakotonirainy M, Arnold S. Development of a new procedure based on the energy charge measurement using ATP bioluminescence assay for the detection of living mold from graphic document. *Luminescence* 2008;23:182-186.

[14] Takahashi T, Nakakita Y, Watari J, Shinotsuka K. Application of a bioluminescence method for the beer industry: sensitivity of MicroStar™-RMDS for detecting beer-spoilage bacteria. *Biosci Biotechnol Biochem* 2000;64(5):1032-1037.

[15] Venkateswaran K, Hattori N, La Duc MT, Kern R. ATP as a biomarker of viable microorganisms in clean-room facilities. *J Microbiol Methods* 2003;52(3):367-377.

[16] Hunter D, Lim D. Rapid detection and identification of bacterial pathogens by using an ATP bioluminescence immunoassay. *J Food Prot* 2010;73(4):739-746.

[17] Kramer M, Suklje-Debeljak H, Kmetec V. Preservative efficacy screening of pharmaceutical formulations using ATP bioluminescence. *Drug Dev Ind Pharm* 2008;34:547-557.

[18] Finger R, Sischo W. Bioluminescence as a technique to evaluate udder preparation. *J Dairy Sci* 2001;84(4):818-823.

[19] Luo J, Liu X, Tian Q, Yue W, Zeng J, Chen G, Cai X. Disposable bioluminescence-based biosensor for detection of bacterial count in food. *Anal Biochem* 2009;394(1):1-6.

[20] Murphy SC, Kozlowski SM, Bandler DK, Boor KJ. Evaluation of adenosine triphosphate-bioluminescence hygiene monitoring for trouble-shooting fluid milk shelf-life problems. *J Dairy Sci* 1998;81:817-820.

[21] Robrish SA, Kemp CW, Bowen WH. Use of extractable adenosine triphosphate to estimate the viable cell mass in dental plaque samples obtained from monkeys. *Appl Environ Microbiol* 1978;35(4):743-749.

[22] Chappelle EW, Levin GV. Use of the firefly bioluminescent reaction for rapid detection and counting of bacteria. *Biochem Med* 1968;2(1);41-52.

[23] Stanley PE. A review of bioluminescent ATP techniques in rapid microbiology. *J Biolumin Chemilumin* 1989;4(1):375-80.

[24] Robrish SA, Kemp CW, Chopp DE, Bowen WH. Viable and total cell masses in dental plaque as measured by bioluminescence methods. *Clin Chem* 1979;25(9):1649-1654.

[25] Moore G, Smyth D, Singleton J, Wilson P. The use of adenosine triphosphate bioluminescence to assess the efficacy of a modified cleaning program implemented within an intensive care setting. *Am J Infect Control* 2010;38(8):617-22.

[26] Forrest WW. Adenosine triphosphate pool during the growth cycle in *Streptococcus faecalis*. *J Bacteriol* 1965;90:1013-8.

[27] Beckers B, Lang HRM, Schimke D, Lammers A. Evaluation of a bioluminescence assay for rapid antimicrobial susceptibility testing of mycobacteria. *Eur J Clin Microbiol* 1985;4(6):556-561.

[28] Willis C, Morley R, Westbury J, Greenwood M, Pallett A. Evaluation of ATP bioluminescence swabbing as a monitoring and training tool for effective hospital cleaning. *British J of Infect Control* 2007;8(5):17-21.

[29] Whiteside TL, Herberman RB. The role of natural killer cells in immune surveillance of cancer. *Curr Opin Immunol* 1995;7:704-710.

[30] Clements MJ, Bommer UA. Translational control: the cancer connection. *Int J Biochem Cell Biol* 1999;31(1):1-23.

[31] Kavanaugh DY, Carbone DP. Immunologic dysfunction in cancer. *Hematol Oncol Clin North Am* 1996;10:927-951.

[32] Devereux TR, Risinger JI, Barrett JC. Mutations and altered expression of the human cancer genes: what they tell us about causes. *IARC Sci Publ* 1999;(146):19-42.

[33] Holt SE, Shay JW (1999). Role of telomerase in cellular proliferation and cancer. *J Cell Physiol* 1999;180:10-18.

[34] Dang CV, Semenza GL. Oncogenic alterations of metabolism. *Trend Biochem Sci* 1999;24:68-72.

[35] Stordal B, Davey M. Understanding cisplatin resistance using cellular models. *IUBMB Life* 1997;59(11): 696–9.

[36] Kangas L, Grönroos M, Nieminen AL. Bioluminescence of cellular ATP: a new method for evaluating cytotoxic agents *in vitro*. *Med Biol* 1984;62(6):338-43.

[37] Garewal, H, Ahmann, F, Schifman, R, Celniker, A. ATP assay: Ability to distinguish cytostatic from cytocidal anticancer drug effects. *J. Natl Cancer Inst.* 1986;77(5):1039-1045.

[38] Contag CH, Jenkins D, Contag PR, Negrin RS. Use of reporter genes for optical measurements of neoplastic disease *in vivo*. *Neoplasia* 2000;2(1-2):41-52.

[39] McMillin DW, Delmore J, Weisberg E, Negri JM, Geer DC, Klippel S, Mitsiades N, Schlossman RL, Munshi NC, Kung AL, Griffin JD, Richardson PG, Anderson KC,

Mitsiades CS. Tumor cell-specific bioluminescence platform to identify stroma-induced changes to anticancer drug activity. *Nat Med* 2010;16(4):483-9. Epub 2010 Mar 14.
[40] Contag CH, Contag PR, Mullins JI, Spilman SD, Stevenson DK, Benaron DA. Photonic detection of bacterial pathogens in living hosts. *Mol Microbial 1995;*18:593-603.
[41] Contag C, Spilman S, Contag P, Oshiro M, Eames B, Dennery P, Stevenson D, Benaron D. Visualizing gene expression in living mammals using a bioluminescent reporter. *Photochem Photobiol 1997;*66:523-531.
[42] Edinger M, Cao YA, Verneris MR, Bachmann MH, Contag CH, Negrin RS. Revealing lymphoma growth and the efficacy of immune cell therapies using in vivo bioluminescence imaging. *Blood* 2003;15;101(2):640-8. Epub 2002 Sep 26.
[43] Rehemtulla A, Stegman LD, Cardozo SJ, Gupta S, Hall DE, Contag CH, Ross BD. Rapid and quantitative assessment of cancer treatment response using in vivo bioluminescence imaging. *Neoplasia* 2000;2(6):491-5.
[44] Negrin RS, Edinger M, Verneris M, Cao YA, Bachmann M, Contag CHH. Visualization of tumor growth and response to NK-T cell based immunotherapy using bioluminescence. *Ann Hematol* 2002;81 Suppl 2:S44-5.
[45] Zhang L, Hellstrom KE, Chen L. Luciferase activity as a marker of tumor burden and as an indicator of tumor response to antineoplastic therapy *in vivo*. *Clin Exp Metastasis* 1994;12:87-92.
[46] Takakuwa K, Fujita K, Kikuchi A, Sugaya S, Yahata T, Aida H, Kurabayashi T, Hasegawa I, Tanaka K. Direct intratumoral gene transfer of the herpes simplex virus thymidine kinase gene with DNA-liposome complexes: Growth inhibition of tumors and lack of localization in normal tissues. *Jpn J Cancer Res* 1997;88: 166-175.
[47] Day RN, Kawecki M, Berry D. Dual-function reporter protein for analysis of gene expression in living cells. *Biotechnology* 1998;25:848-850.
[48] Grentzmann G, Ingram JA, Kelly PJ, Gesteland RF, Atkins JF. A dual-luciferase reporter system for studying recording signals. *RNA* 1998;4:479-486.
[49] Weissleder R, Tung CH, Mahmood U, Bogdanov A, JR (1999). *In vivo* imaging of tumor with protease-activated near-infrared fluorescent probes. *Nat Biotechnol* 1999;17:375-378.
[50] Reynolds JS, Troy TL, Mayer RH, Thompson AB, Waters DJ, Cornell KK, Snyder PW, Sevick-Muraca EM. Imaging of spontaneous canine mammary tumors using fluorescent contrast agents. *Photochem Photobiol* 1999;70(1):87-94.
[51] Nioka S, Yung Y, Shnall M, Zhao S, Orel S, Xie C, Chance B, Solin L. Optical imaging of breast tumor by means of continuous waves. *Adv Exp Med Biol* 1997;411:227-32.
[52] Chance B, Luo Q, Nioka S, Alsop DC, Detre JA. Optical investigations of physiology: a study of intrinsic and extrinsic biomedical contrast. *Philos Trans R Soc London Biol Sci* 1997;352(1354):707-716.
[53] Contag PR, Olomu IN, Stevenson DK, Contag CH. Bioluminescent indicators in living mammals. *Nat Med* 1998;4(2):245-247.
[54] Edinger M, Sweeney T, Mailander V, Tucker A, Olomu AB, Negrin RS, Contag CH. Non-invasive assessment of tumor cell proliferation in animal models. *Neoplasia* 1999;1(4):303-310.
[55] Thompson EM, Adenot P, Tsuji FI, Renard JP. Real-time imaging of transcriptional activity in live mouse preimplantation embryos using a secreted luciferase. *Proc Natl Acad Sci USA* 1995;92(5):1317-1321.

[56] Willard ST, Faught WJ, Frawley LS. Real-time monitoring of estrogen-regulated gene expression in single, living breast cancer cells: a new paradigm for the study of molecular dynamics. *Cancer Res* 1997;57(20):4447-4450.
[57] Chen H, Biel MA, Borges MW, Thiagalingam A, Nelkin BD, Baylin SB, Ball DW. Tissue-specific expression of human achaete-scute homologue-1 in neuroendocrine tumors: transcriptional regulation by dual inhibitory regions. *Cell Growth Differ* 1997;8(6):677-686.
[58] Pettersson J, Nordfelth R, Dubinina E, Bergman T, Gustafsson M, Magnusson KE, Wof-Watz H. Modulation of virulence factor expression by pathogen target cell contact. *Science* 1996;273(5279):1231-1233.
[59] Hooper CE, Ansorge RE, Rushbrooke, JG. Low-light imaging technology in the life sciences. *J Biolumin Chemilumin* 1994;9(3):113-122.
[60] Hooper CE, Ansorge RE, Brown HM, Tomkins P. CCD imaging of luciferase gene expression in single mammalian cells. *J Biolumin Chemilumin* 1990;5(2):123-130 (1990).
[61] White MR, Masuko M, Amet L, Elliott G, Braddock M, Kingsman AJ, Kingsman SM. Real-time analysis of the transcriptional regulation of HIV and hCMV promoters in single mammalian cells. *J Cell Sci* 1995;108 (Pt 2):441-455.
[62] Karsi A, Gülsoy N, Corb E, Dumpala PR, Lawrence ML. High-throughput bioluminescence-based mutant screening strategy for identification of bacterial virulence genes. *Appl Environ Microbiol* 2009;75(7):2166-75.
[63] Ton That V, Nguyen S, Poon D, Monahan WS, Sauerwein R, Lafferty DC, Teasdale LM, Rice AL, Carter W, Maier T, Machida CA. Bioluminescent *lux* gene biosensors in oral streptococci: determination of complementary antimicrobial activity of minocycline hydrochloride with the anesthetic lidocaine/prilocaine or the antiseptic chlorhexidine. 2010 In: *Periodontitis: Symptoms, Treatment and Prevention; Public Health in the 21st Century*. Nova Science Publishers.
[64] Hutchens M, Luker GD. Applications of bioluminescence imaging to the study of infectious diseases. *Cell Microbiol* 2007;9(10):2315-22.
[65] Kumar V, Selvakumar N, Venkatesan P, Chandrasekaran V, Paramasivan CN, Prabhakar R. Bioluminescence assay of adenosine triphosphate in drug susceptibility testing of *Mycobacterium tuberculosis*. *Indian J Med Res* 1998;107:75-7.
[66] www.cdc.gov
[67] Bowden GH. Does assessment of microbial composition of plaque/saliva allow for diagnosis of disease activity of individuals? *Community Dent Oral Epidemiol* 1997;25:76-81.
[68] Van Houte J. Microbiological predictors of caries risk. *Adv Dent Res* 1993;7(2):87-96.
[69] Carounanidy U, Sathyanarayanan R. Dental caries: A complete changeover (Part II) – Changeover in the diagnosis and prognosis. *J Conserv Dent* 2009;12(3):87-100.
[70] Balakrishnan M, Simmonds RS, Tagg JR. Dental caries is a preventable infectious disease. *Australian Dental Journal* 2000;45(4):235-245.
[71] Hojo K, Nagakota S, Ohshima T, Maeda N. Bacterial interactions in dental biofilm development. *J Dent Res* 2009;88:982.
[72] Azrak B, Callaway A, Willerhausen B, Ebadi S, Gleissner C. Comparison of a new chairside test for caries risk assessment with established methods in children. *Schweiz Monatsschr Zahnmed* 2008;118(8):702-8.

[73] Rupf S, Merte K, Eschrich K. Quantification of bacteria in oral samples by competitive polymerase chain reaction. *J Dent Res* 1999;78:850-856.
[74] Rupf S, Kneist S, Merte K, Eschrich K. Quantitative determination of *Streptococcus mutans* by competitive polymerase chain reaction. *Eur J Oral Sci* 1999;167:75-81.
[75] Ronner P, Friel E, Czerniawski K, Frankle S. Luminometric assays of ATP, phosphocreatine, and creatine for estimation of free ADP and AMP. *Analytical Biochemistry* 1999;275:208-216.
[76] Kemp CW. Adenylate energy charge: A method for the determination of viable cell mass in dental plaque samples. *J Dent Res* 1979;68(D):2192-2197.
[77] Stecksen-Blicks C. Salivary counts of lactobacilli and *Streptococcus mutans* in caries prediction. *Scand J Dent Res* 1985;93(3):204-212.
[78] Crossner CG, Unell L. Salivary lactobacillus counts as a diagnostic and didactic tool in caries prevention. *Community Dent Oral Epidemiol* 1986;14(3):156-160.
[79] Crossner CG. Salivary lactobacillus counts in the prediction of caries activity. *Community Dent Oral Epidemiol* 1981;9(4):182-190.
[80] Mundorff SA, Eisenberg AD, Leverett DH, Espeland MA, Proskin HM. Correlations between numbers of microflora in plaque and saliva. *Caries Res* 1990;24(5):312-317.
[81] Loesche, WJ. Role of *Streptococcus mutans* in human dental decay. *Microbiol Rev* 1986;50(4):353-380.
[82] Russell MW, Lehner T. Characterization of antigens extracted from cells and culture fluids of *Streptococcus mutans* serotype c. *Arch Oral Biol* 1978;23(1):7–15.
[83] Matsumura M, Izumi T, Matsumoto M, Tsuji M, Fujiwara T, Ooshima T. The role of glucan-binding proteins in the cariogenicity of *Streptococcus mutans*. *Microbiol Immunol* 2003;7(3):213–15.
[84] Gibbons RJ, Cohen L, Hay DI. Strains of *Streptococcus mutans* and *Streptococcus sobrinus* attach to different pellicle receptors. *Infect Immun* 1986;52:555-56.
[85] Arends J, Christoffersen J. The nature of early caries lesions in enamel. *J Dent Res* 1986;65:2–11.
[86] Okada M, Soda Y, Hayashi F, Doi T, Suzuki J, Miura K, Kozai K. Longitudinal studies on dental caries incidence associated with *Streptococcus mutans* and *Streptococcus sobrinus* in pre-school children. *J Med Microbiol* 2005;54:661–65.
[87] Coykendall AL. Classification and identification of the Viridans streptococci. *Clinical Microbiology Reviews* 1989;2(3):315–28.
[88] Wood WA. Fermentation of carbohydrates and related compounds. In: Gunsalus IC, Stanier, RY, eds. *The bacteria: treatise on structure and function*. Vol. 2. New York: Academic Press, 1961:59-149.
[89] Carlsson J, Grahnen H, Jonsson G. Lactobacilli and streptococci in the mouth of children. *Caries Res* 1975;9:333-9.
[90] Babaahmady KG, Challacombe SJ, Marsh PD, Newman HN. Ecological study of *Streptococcus mutans*, *Streptococcus sobrinus* and *Lactobacillus* spp. at sub-sites from approximal dental plaque from children. *Caries Res* 1998;32(1):51-8.
[91] Van Houte J, Gibbons RJ, Pulkkinen AJ. Ecology of human oral lactobacilli. *Infect Immun* 1972;6:723-9.
[92] Loesche WJ, Eklund S, Earnest R, Burt B. Longitudinal investigation of bacteriology of human fissure decay: epidemiological studies in molars shortly after eruption. *Infect Immun* 1984;46:765-72.

[93] Grindefjord M, Dahllof G, Nilsson B, Modeer T. Prediction of dental caries development in 1-year-old children. *Caries Res* 1995;29:343-8.
[94] Ravald N, Hamp SE, Birkhed D. Long-term evaluation of root surface caries in periodontally treated patients. *J Clin Periodontol* 1986;13:758-67.
[95] Goodman AH, Armelagos GH. Factors affecting the distribution of enamel hypoplasias within the human permanent dentition. *American Journal of Physical Anthropology* 1985;68:479-493.
[96] [My paper]Dummer PM, Kingdon A, Kingdon R. Distribution of developmental defects of tooth enamel by tooth-type in 11-12-year-old children in South Wales. *Community Dent Oral Epidemiol* 1986;14(6):341-4.
[97] Li Y, Navia JM, Bian JY. Prevalence and distribution of developmental enamel defects in primary dentition of Chinese children 3-5 years old. *Community Dent Oral Epidemiol* 1995;23(2):72-9.
[98] Alaluusua S, Malmivirta R. Early plaque accumulation--a sign for caries risk in young children. *Community Dent Oral Epidemiology* 1994; 22(5 Pt 1):273-6.
[99] Wendt LK, Hallonsten AL, Koch G, Birhed D. Analysis of caries-related factors in infants and toddlers living in Sweden. *Acta Odontologica Scandinavica* 1996;54:131–137.
[100] Lindhe J, Axelsson P, Tollskog G. Effect of proper oral hygiene in gingivitis and dental caries in Swedish school children. *Community Dent Oral Epidemiol* 1975;3:150-155.
[101] Poulsen S, Agerbaek N, Melson B, Glavind L, Rolla G. The effect of professional toothcleansing on gingivitis and dental caries in children after 1 year. *Community Dent Oral Epidemiol* 1976;4:195-199.
[102] Leverett DH, Featherstone JDB, Proskin HM, Adair SM, Eisenberg AD, Mundorffshrestha SA, Shields CP, Shaffer CL, Billings RJ. Caries risk assessment by a cross-sectional discrimination model. *J Dent Res* 1993;72:529-537.
[103] Zero D, Fontana M, Lennon AM. Clinical applications and outcomes of using indicators of risk in caries management. *J Dent Edu* 2001;65(10):1126-32.
[104] Li Y, Wang W. Predicting caries in permanent teeth from caries in primary teeth: An eight-year cohort study. *J Dent Res* 2002;81:561-6.
[105] Greenwell AL, Johnsen D, DiSantis TA, Gerstenmaier J, Limbert N. Longitudinal evaluation of caries patterns from the primary to the mixed dentition. *Pediatr Dent* 1990;12:278-282.
[106] Loesche WJ, Rowan J, Straffon LH, Loos PJ. Association of *Streptococcus mutans* with human dental decay. *Infect Immun* 1975;11:1252-1260.
[107] Krasse B. Biological factors as indicators of future caries. *Int Dent J* 1988;38:219-225.
[108] Littleton NW, Kakehashi S, Fitzgerald RJ. Recovery of specific "caries-inducing" streptococci from carious lesions in the teeth of children. *Arch Oral Biol* 1970;15:461-463.
[109] Klock B, Krasse B. Microbial and salivary conditions in 9- to 12-year-old children. *Scand J Dent Res* 1977;85:56-63.
[110] Zickert I, Emilson CG, Krasse B. Correlation of level of duration of *Streptococcus mutans* infection with incidence of dental caries. *Infect Immun* 1983;39:982-985.
[111] Newbrun E, Matsukubo T, Hoover CI, Graves RC, Brown AT, Disney JA, Bohannan HM. Comparison of two screening tests for *Streptococcus mutans* and evaluation of

their suitability for mass screenings and private practice. *Community Dent Oral Epidemiol* 1984;12:325-331.

[112] Kristoffersson K, Axelsson P, Birkhed D, Bratthall D. Caries prevalence, salivary *Streptococcus mutans* and dietary scores in 13-year-old Swedish schoolchildren. *Community Dent Oral Epidemiol* 1986;14:202-205.

[113] Alaluusua S, Kleemol-Kujala E, Nystrom M, Evalahti M, Gronroos L. Caries in the primary teeth and salivary *Streptococcus mutans* and lactobacillus levels as indicators of caries in permanent teeth. *Pediatr Dent* 1987;9:126-130.

[114] Boyar RM, Bowden GH. The microflora associated with the progression of incipient carious lesions in teeth of children living in a water-fluoridated area. *Caries Res* 1985;19:298-306.

[115] Lang NP, Hotz PR, Gusberti FA, Joss A. Longitudinal clinical and microbiological study on the relationship between infection with *Streptococcus mutans* and the development of caries in humans. *Oral Microbiol Immunol* 1987;2:39-47.

[116] Kingman A, Little W, Gomez I, Heifetz SB, Driscoll WS, Sheats R, Supan P. Salivary levels of *Streptococcus mutans* and lactobacilli and dental caries experiences in a US adolescent population. *Community Dent Oral Epidemiol* 1988;16:98-103.

[117] Mundorff SA, Billings RJ, Leverett DH, Featherstone JD, Gwinner LM, Shields CP, Proskin HM, Shaffer CL. *Saliva and dental caries risk assessment.* Ann NY Acad Sci 1993;694:302-4.

[118] Anderson MH, Shi W. A probiotic approach to caries management. *Pediatr Dent* 2006;28(2):151-3; discussion 192-8.

[119] American Academy of Pediatrics: Policy Statement. Organizational principles to guide and define the child health care system and/or improve the health of all children. Section on Pediatric Dentistry. Oral Health Risk Assessment Timing and Establishment of the Dental Home. *Pediatrics* 2003;111(5):1113-1116.

Reviewed by: Dr. Elizabeth Palmer, Assistant Professor, Department of Pediatric Dentistry, OHSU School of Dentistry and Dr. John Peterson, Professor, Department of Pediatric Dentistry, OHSU School of Dentistry.

In: Bioluminescence
Editor: David J. Rodgerson, pp. 29-48

ISBN 978-1-61209-747-3

Chapter 2

NEW METHODOLOGIES BASED ON THE COUPLING OF LUCIFERASE WITH NANOMATERIALS

Simone M. Marques and Joaquim C. G. Esteves da Silva [*]
Centro de Investigação em Química, Department of Chemistry and Biochemistry, Faculty of Sciences, Universidade do Porto, Rua do Campo Alegre 687, 4169-007 Porto, Portugal

ABSTRACT

Luciferase is a powerful tool in bioanalysis. Several well-established methods employ luciferases, particularly firefly and *Renilla* luciferases, as reporter genes or biosensors in environmental, biomedical and biochemical research. These techniques have interesting features for the analyst such as sensitivity, specificity and reduced assay time. Nanochemistry and Nanotechnology are disciplines that are gaining much attention and evolving rapidly. They allow the development of custom-made nanomaterials with the desired properties, starting from conventional bulk materials. Recently, the coupling of nanomaterials such as carbon nanotubes, mesoporous silica nanoparticles, metallic nanoparticles and quantum dots with luciferases led to new or improved methodologies for analyte quantification and enhanced gene delivery strategies. One of the principal scopes is to modulate or alter luciferase's bioluminescence emission, either by stabilizing it or tuning it to longer wavelengths. This chapter aims to present state-of-art articles regarding new methods based on the coupling of luciferases to nanomaterials, along with a brief introduction to Nanoscience.

Keywords: Luciferase; Nanochemistry; Nanomaterials; Nanoscience; Nanotechnology; Bioanalytical Chemistry; Bioimaging; Biomedicine; Carbon Nanotubes; Gold Nanoparticles; Nanodiamonds; Nanostructured Films; Quantum Dots; Silica Nanoparticles

[*] Tel: +351220402569. Fax: +351220402659. E-mail: jcsilva@fc.up.pt

INTRODUCTION

Basic Concepts about Nanoscience, Nanomaterials, Nanochemistry and Nanotechnology

Nanoscience is a current, although not really novel [1], topic of intense interest and, accordingly, of much research. The very first definition of Nanoscience was strictly based on size; any material within 1-100 nm would be under the remit of Nanoscience. Today, however, the scientific community does not hold this single definition. In fact, in 2004 the Royal Society and the Royal Academy of Engineering published a report in which a broader definition was proposed [2]:

> "Nanoscience is the study of phenomena and manipulation of materials at atomic, molecular and macromolecular scales, where properties differ significantly from those at a larger scale."

This definition highlights the concept of size-dependent properties of a given material. A nanomaterial has the same composition of the macroscopic, or bulk, material, but a whole different set of properties such as spectroscopic, mechanical, chemical reactivity, among others. Usually this shift in properties occurs at the nanometric scale, and then the "nano" designation. However, some materials already present bulk properties at a few nanometers in size, while others can be regarded as nanomaterials in the micrometer range, and that is one reason why the "size only" definition is not an ideal one and that an extended scale from 1 to 1000 nm is more realistic. Furthermore, at this scale, a defined structure can be observed with spectroscopic techniques in true nanomaterials or nanostructured materials (bulk materiais with defined forms at nanoscale, as in the case of zeolites), but not in bulk materials in general. For example, pure bulk metallic gold is a yellow and inert material whereas colloidal gold nanoparticles not only exhibit several different colours according to size and concentration but also present catalytic properties [3], as in the oxidation of carbon monoxide to carbon dioxide [4]. Beyond nanogold, the semiconductor nanoparticles known as "quantum dots" have receiving much attention due to their intense and stable fluorescence and wide variety of colours. These unique properties result from a quantic confinement of the nanocrystal not found in bulk semiconductors [5, 6].

Nanomaterials include, besides nanomaterials themselves and nanostructured materials already mentioned, the nanocomposites and nanohybrids. Nanocomposites are a combination of at least two different components into a single nanomaterial with final properties that can be the same of the original individual components or new ones can emerge. For example, the integration of zinc oxide nanoparticles in plain cellulosic paper led to an improved strain sensor [7]. By its turn, a nanohybrid is the result of coupling different nanomaterials in which their physical integrity is maintained in the final product and new or improved features are obtained [8]. As an example, the conjugation of single-walled carbon nanotubes with quantum dots not only allowed the study of photochemical processes in which they participate but also revealed their potential application in photovoltaic cells [9]. Another important attribute of nanomaterials is their high surface area-volume ratio, which is in the order of 100 to 1000 m^2/g. For this reason, mesoporous nano-sized silica is now being preferred over bulk silica with the same porosity due to the higher superficial area of the nanomaterial.

In the nanoworld, surface is a very important aspect. In bulk materials, surfaces are a minor portion of their volume and can be ignored up to some point. But almost all of a nanomaterial is surface. To obtain a nanomaterial its size has to be strictly controlled, to avoid it becoming bulk. As a result of this control, an incomplete filling of the valence orbitals occurs. These "dangling bonds" raise the surface's energy and explain the extreme reactivity of nanomaterials' surfaces. This reactivity though is not only detrimental; by adding capping agents not only this energy is reduced and more stability is obtained but also the functionalization of the nanomaterial is achieved. Functionalization is the attachment of molecules through covalent or non-covalent (electrostatic, physical adsorption) interactions which confers new and improved characteristics to the nanomaterial. For example, the functionalization of single-walled carbon nanotube sensors with benzene enhanced their sensitivity towards gaseous species derived from sulphur hexafluoride [10].

Nanomaterials can be obtained by chemical or physical process. In this context, Nanochemistry can be regarded as "the utilization of synthetic chemistry to make nanoscale building blocks of different size and shape, composition and surface structure, charge and functionality" [11]. It is the contribution of Chemistry to Nanoscience. Traditionally, Chemistry deals with molecules, whose dimensions are inferior to nanometers, and so to obtain nanomaterials one needs to build it piece by piece, an approach known as "bottom-up". In Physics, on the other hand, one begins with "large" blocks of bulk material, in the micrometer range for example, which are then modelled to the desired nanomaterial towards miniaturization, a "top-down" approach compared to the sculpture of a statue. This division is not absolute, however, as top-down techniques can also be applied in the chemical preparation of nanomaterials. For instance, the exfoliation of clay results in separated lamellar sheets with different properties from the bulk clay and nanometric dimensions. It is also important to note that the real innovation is neither in the building block composition nor in the synthesis process but rather in the manner they arrange. In Nanoscience an important process is the self-assembly, regarded as the spontaneous generation of higher-order systems from pre-existing components that already contain the information for their self-assembly, leading to the synthesis of nanomaterials with new or betters properties in relation to the starting building blocks [12, 13]. In Nature, the phenomenon of self-assembly occurs in scales from the folding of proteins, for example, to the existence of galaxies. Self-assembly is also a reversible process, implying the adjustment of the interactions according to novel conditions. In Nanochemistry's bottom up methods, the most popular self-assembly process is the molecular self-assembly of atoms, molecules or ions into bigger and ordered structures based in commonly known chemical interactions such as ionic, metallic, covalent, electrostatic, hydrogen bridges, Van der Waals forces and π-π stacking. However, self-assembly is not limited to small molecules aggregates. Non-molecular self-assembly leads to even bigger species, from nano to the micrometer range. In this case other forces come into play, like capillary effects, Van der Waals and London forces and elastic, electric and magnetic interactions, that is, forces that do not hold for individual molecules but requires cooperativity among them. In equilibrium, and without external forces, self-assembly occurs by minimizing the system's energy, the so-called static self-assembly. Once formed, the system is stable and it is at global or local energy minima [12, 13]. On the other hand, out of equilibrium and in the presence of external forces applied to the system, it undergoes a dynamic self-assembly

[14]. Usually, these forces are mild, so that in their absence the system does not disaggregate. As the research has focused on static self-assembly much is known about it [12, 13]. In contrast, dynamic self-assembly demands more studies about its theoretical basis and practical applications [14].

In Nanoscience research, one theoretically designs a certain nanomaterial with already defined characteristics, like enhanced luminescence or mechanical resistence, which is then fabricated or synthesized and characterized by appropriate techniques and so, unlike other disciplines, fundamental science is in mutual association with experimental science. In fact Nanoscience and Nanotechnology evolved almost at the same time, having existed a delay in Nanotechnology outcome just due to the need of improved instrumental devices. The big challenge is, so, to attribute practical use to them, creating new devices, structures and systems for general use through commercial distribution and industrial scale-up.

Why Nanomaterials-Luciferase Coupling?

Luciferase is a key tool in biomedical and bioanalytical chemistry [15]. Many efforts are made towards improving its features, namely increased catalytic efficiency [16], higher luminescence intensity and stability [17-19], new light colours emission [20, 21], and enhanced resistance to proteolysis [22] and to elevated temperatures and pH variations [23]. So far, these features were mainly accomplished through genetic engineering, mainly by mutagenesis in selected amino acid [16-20], by fusion with other enzymes [24] or biomolecules [25] and by splitting luciferase into two or more fragments which regain activity through protein-protein interactions between the proteins attached to each of the luciferase's portions [26, 27]. In some cases, other improvements were obtained through the use of biological or chemical adjuvants which can enhance the biochemical reaction and the corresponding light output, for example α-synuclein [28], inorganic pyrophosphate [29], coenzyme A [30], liposomes [31, 32], magnesium sulfate [33] and other osmolites [34]. Chemically modified luciferins also confer distinctive patterns to the bioluminescent reaction [35, 36].

Regarding Nanoscience and the respective nanomaterials and nanodevices, it soon became evident their potential in biological applications [37]. Bionanotechnology encompass the study, characterization and application of nanomaterials in living entities. Some applications already proposed include photothermal therapy for cancer [38], novel vectors for gene delivery [39] and novel probes for bioimaging [40], just to quote a few.

Following these ideas a logical step would be to bring luciferase and nanomaterials together as nanohybrids, improving luciferase's properties and expanding its range of applications, as will now be exposed.

New Methodologies Based on the Coupling of Luciferase with Nanomaterials

Carbon-Based Nanomaterials

Carbon Nanotubes

Carbon is an eclectic element concerning nanomaterial production. It appears in several forms:

i) graphene nanoribbons, a one atom-thick sheet of conjugated sp^2-hybridized carbon atoms [41- 44];
ii) filamentous carbon (nanotubes if hollow, nanofibers if filled), which can be regarded as a graphene sheet rolled around itself to form a cylinder with 0.4-3 nm in diameter if composed of only one sheet (a single-walled carbon nanotube), or 2-100 nm in diameter if composed of several concentric sheets (a multi-walled carbon nanotube), and lengths from nanometers up to micrometers or even several centimetres [41, 44, 45];
iii) fullerenes, spherical C_{60} or higher [41, 46];
iv) nanodiamonds [41, 47]; and
v) nanostructured porous carbon replicas, structures based on the utilization of templates, commonly mesoporous silica, where the self-assembly of carbon occurs; posteriorly, the template is removed and the material will present the inverse of the shape of the template and a specific pore size [48-50].

Carbon nanotubes are a particularly successful carbon-based nanomaterial. They present very attractive characteristics like high mechanical strength, excellent chemical, electronic and optoelectronic performances, low density, and good heat transmission capacity [41, 44, 45]. Another important feature is the great diversity of functionalization processes they allow, especially at the edges and defects. There is also the possibility of polymer wrapping and endohedral encapsulation of molecules by inserting molecular components inside them and allowing their self-assembly [51-54]. These processes are exclusive of nanotubes [51-54]. Taken together with the high surface areas they possess, the possibilities are astonishing. Those improvements have rendered several interesting applications for carbon nanotubes as catalysts [55], in electric circuits and devices [56], in solar cells [9], as sensors [10], among others. Biomedicine gives much attention to the evolution of these nanomaterials [44, 57, 58], and current studies are focused on their biocompatibility [59] and improving their solubility in organic solvents and water [60].

A biosensor for cellular ATP based on carbon nanotubes and luciferase was proposed [61]. It relies upon the immobilization of luciferase into near-infrared fluorescence emitting nanotubes through carboxylated poly-ethylene glycol phospholipids wrapped around commercial single-walled nanotubes. When the bioluminescent reaction catalyzed by luciferase takes place, with concomitant consume of cellular ATP and exogenously-added luciferin, the resulting product, oxyluciferin, will adhere to the nanotubes and quench their fluorescent emission. It allows the quantification of ATP in roughly 10 minutes at room temperature, it is sensitive and selective to ATP, as it does not respond neither to closely related nucleotides as ADP, CTP and GTP nor to the bioluminescent reaction by-products AMP and inorganic pyrophosphate, and it allows the spatial and temporal quantification of ATP in living cells. On the other hand, it is an irreversible sensor that cannot be regenerated after the bioluminescent reaction takes place. With further optimizations this could be a useful optical biosensor for ATP. The principal contribution of the nanomaterial in relation to previously proposed methods is the enhancement in sensitivity, simplicity and stability that they confer to the biosensor [61].

Nanodiamonds

Nanodiamonds are allotropes of carbon, with an average diameter of 5 nm, formed from the detonation of carbon-bearing explosives such as a mixture of trinitrotoluene (TNT) and hexogen, and so they are also called detonation nanodiamonds or ultrafine-dispersed diamonds (UDD) [41, 47]. They possess interesting characteristic, namely a large surface area, optical transparency and luminescence, increased mechanical strength, and enhanced magnetic and electrochemical properties [41, 47, 62]. Their core is composed of crystalline carbon with some nitrogen atoms derived from the precursor that, together with a nearby lattice vacancy can be regarded as defects responsible for their photoluminescence [47, 63, 64]. Nanodiamonds are already produced in commercial scale, for example by ALIT, Inc., at Kiev (Ukraine) [65]. Recently, however, they are being recruited to biomedical applications as drugs, genes and proteins carriers [66, 67], as a cellular scaffold [68], as a bioimaging probe [69] and as biosensors [70, 71], provided that they are non-cytotoxic; in fact, they are considered the most biocompatible of all carbon nanomaterials [47], and more applications are likely to be proposed as they are prone to form nanocomposites and nanohybrid materials and to be subjected to further functionalizations [56].

A proof-of-principle bioluminescent sensor based on nanodiamonds and bacterial luciferase is described in the literature [72]. A biochip was developed by adsorbing the nanodiamod-luciferase complex into an aluminium oxide film, and a bioluminescent signal was detected. Although this is a preliminary study it points out that the nanodiamonds-luciferase coupling holds potential for bioanalytical applications.

Metallic Nanoparticles

Metals in their nanometric scale are among the first materials to get practical applications and posteriorly recognized as materials different from bulk with size-depending properties [1, 73]. This kind of nanomaterial can be obtained in a variety of shapes and sizes [74-76] and their interesting properties grant them lots of studies and proposed applications. The most common metallic nanoparticles are made from noble metals like gold, silver, platinum and palladium. In the bulk form they are little reactive, but they became reactive in nano form [4, 73]. On the other hand, metals which are already reactive in the bulk form become so reactive as nanometals that are very difficult to work with them. The principal applications are based on their optical, magnetic and catalytic properties [73]. For example, the use of silver nanoprisms in solar cells enhanced their light harvesting potential [77]. In biological applications gold nanoparticles is undoubtedly one of those that receive plenty of attention and studies [78-80].

Gold Nanoparticles

The nanochemistry of gold nanoparticles, or colloidal gold, has old roots, being used since ancient times for decoration and as a medicine, for example [1, 81]. Gold nanoparticles are formed by metallic gold with dangling bonds at the surface, like other nanomaterials. The most distinctive characteristic of their surface chemistry is the high affinity for thiol groups. This leads to the formation of self-assembled monolayers (SAM), especially of alkanethiols, which stabilized the surface atoms and confers new functionalities to the surface [82]. Furthermore, it is possible the synthesis of hybrid nanoparticles, like a silver core covered by a gold layer [83], gold nanoparticles conjugate with platinum clusters at their surface [84] or the trapping of iron oxide inside gold nanoparticles [85], which enhances their features and augment the field of applications.

In metallic surfaces, the existence of free electrons in movement generates an electric dipole at the surfaces, which induces their collective oscillation, the so-called plasmons [86]. By irradiating the particles with light at a specific wavelength, they can resonate, giving rise to absorption of the radiation and detection of a plasmonic band, responsible for the intense color of gold nanoparticles. The frequency at which this plasmonic band occurs depends on the shape and size of the nanoparticles, but for gold it is in the visible to near-infrared range. For instance, in a recent study, oligonucleotides were immobilized onto gold nanoparticles [87]. By shedding light at the plasmonic frequency, the light was absorbed and led to an electronic transition from ground state to an excited state. The non-radiative decay to the fundamental state released energy as heat. The heat so generated denatured the oligonucleotide chains which then exerted their intended activity in this study, gene silencing [87]. In phototherapy, nanoparticles could be directed to cancer cells [88] or bacteria [89]; their exposition to near-infrared emitting gold nanoparticles causes cellular lysis. Gold nanoparticles also present magnetic properties, which could be applied for data storage [90].

Many achievements could be obtained with gold nanoparticles' functionalization with biomolecules. For example, the coupling of the enzyme tyrosinase onto gold nanoparticles led to a biosensor for phenolic compounds [91]. A biosensor for proteases based on luciferase-gold nanoparticles is already described [92]. It is based on the quenching, by the nanoparticles, of the light emitted by luciferase. A blue-emitting, eight-mutation variant with enhanced stability in serum and higher catalytic efficiency luciferase from *Renilla reniformis*, termed Luc8 [17], was coupled to 5 nm-diameter gold nanoparticles previously functionalized with carboxylic oligo(ethylene glycol) through a short amino acid sequence recognized by matrix metalloproteinase-2 (MMP-2), which was chosen as a test protease due to its relevant cellular roles, namely in tumor invasion and metastasis [93]. By acting on its substrate, MMP-2 will promote the separation between luciferase and the nanoparticles, thus raising the light emission [92]. Protease concentrations ranging from 50 ng mL^{-1} to 1 μg mL^{-1} were assessed. This example represents an easy and sensitive way to determine MMP-2 concentration, but other proteases could be assayed by just changing the amino acid sequence. Furthermore, it relies upon desired features conferred by the nanoparticles such as photostability, biocompatibility and large surface area to attach luciferase, which are not easily obtained with traditional chemical quenchers.

Quantum Dots

Quantum dots (QD) is the common designation of semiconductor colloidal nanocrystals, generally measuring 2-6 nm, composed of elements from the periodic groups II and VI (e.g. CdSe, CdTe, CdS, ZnSe) or III and V (e.g. InP, InAs), sometimes with a core-shell structure like CdSe/ZnS or CdTe/CdS or dopped with another compound, such as Mn: ZnS/ZnS [94]. Recently, the production of carbon-dots, the carbon-based counterpart of inorganic QDs [95, 96], is another bid on QDs' versatility. They present several unique optical and electronic features, like size- and composition-tunable light emission from visible to near-infrared wavelengths, broad absorption spectra, narrow and symmetric emission spectra, very high levels of brightness and photostability, high quantum yields and high molar extinction coefficients [94], being one of the most popular nanomaterial to date.

QDs have so many applications that refer all of them is beyond the scope of this chapter. Illustrative examples include latent fingerprinting detection [97, 98], biosensor for hydrogen peroxide [99], and artificial light harvester in solar energy systems [100]. But it is in the biomedical field that QDs really stand out [40, 101-105]. They are applied as probes for *in vitro* and *in vivo* imaging [106-108], in real-time biomolecule tracking [109-111], as biosensors for kinases and phosphatases [112] and glucose [113], and so on. However, for these biological applications, the issue of biocompatibility and toxicity is always under attention. Current studies suggest that QDs can insert cellular damages [114-118], but deleterious effects can be diminished through capping with an additional shell, like in the core/shell QDs [108], functionalization with ligands [115, 119] and water-soluble QDs formulation [107]. Another strategy is the synthesis of QDs devoided of toxic elements, especially cadmium [103, 108, 120, 121].

Bioimaging is observing an intense development. Application of luciferases as probes are desired due to simplicity, costs and versatility. However, a major drawback is encountered: the way light interacts with tissues. In biological systems, chromophores molecules like hemoglobin, colagen and water could absorb light in certain wavelengths, which are then transmitted or scattered [122]. In this way little light can penetrate more than a few inches, except for red or near-infrared radiation [122]. Green- or blue-emitting luciferase, like those from fireflies and *Renilla*, are inadequate for such *in vivo* bioimaging studies. The cloning of red light-emitting luciferases or the genetic engineering of the green-emitting ones was a possible solution [20, 21]. Recently, however, a novel proposal was made, based on the bioluminescence resonance energy transfer (BRET) from the eight-mutation luciferase Luc8 [17] to a red-emitting quantum dot [123]. Resonant energy transfer is a long-recognized phenomenon. A theory to explain it was proposed by the German scientist Theodor Förster, and hence the acronym Förster Resonant Energy Transfer, FRET, commonly used [124-126]. It involves two chromophores, a donor and an acceptor. The donor, initially in an electronically excited state, can transfer the energy to an acceptor as long as they are close to each other, between 1-10 nm, and that the donor's emission spectrum overlaps the acceptor's absorption spectrum. This transfer occurs through dipole-dipole coupling, and not by electron transfer, whereby it is non-radiative, or resonant [125, 126]. When both the chromophores are fluorescent the term Fluorescent Resonance Energy Transfer applies. If the energy comes from a bioluminescent reaction it is called Bioluminescence Resonance Energy Transfer [127-129]. There are numerous examples of FRET involving QDs and fluorophores [130], fluorescent proteins [131] or nanomaterials [132].

In the proposed system [123], Luc8 (emission maximum at 480 nm) acted as the donor molecule and commercial red-emitting QDs (CdSe/ZnS core-shell QDs with emission peak at 605 and 655 nm, and CdTe/ZnS at 705 and 800 nm) were the acceptors. These QDs have a maximum absorption in the blue range, hence a good energy transfer would be expected.

The QD-Luc8 conjugates were initially characterized regarding coupling efficiency and effective BRET occurrence. By adding coelenterazine, a peak at 480 nm was indeed registered along with a second peak at 605-800 nm due to BRET. Then they were injected in solution in nude mice and the BRET signal was detected with a proper camera. It was verified that not only BRET occurred *in vivo* but also deeper intramuscular injections allowed a recordable signal from only 5 pmol of conjugates, compared to little signal registered for 30 pmol of bare luciferase bioluminescence at the same locations or at subcutaneous injections. The same animals were tested for QD-Luc8 fluorescent emission and, although a signal could be obtained in subcutaneous injections, as expected, it was very faint in intramuscular injections, confirming that bioluminescence detection is more sensitive than fluorescence *in vivo*. At the same study, nude mice were injected simultaneously with QD-Luc8 conjugates with different emission maxima, from 605 to 800 nm, and the corresponding signals were sequentially detected by using adequate filters to each wavelength, thus confirming the possibility of multiplex imaging. Finally it was tested the possibility of monitoring cells transfected with those conjugates *in vivo*, instead of injecting the conjugates in solution or buffers. To achieve it, C6 glioma cells were incubated *in vitro* with QD655-Luc8 conjugates and posteriorly injected in nude mice. A strong BRET signal was obtained both *in vitro* and *in vivo* from these cells, but no fluorescence from the QDs was detected.

In subsequent studies the system was improved regarding the QD-Luc8 coupling, by using HaloTag- [133] and intein-mediated conjugation [134], and enhancing the long-term stability of Luc8 by encapsulation in a polyacrylamide gel [135]. Another improvement is the luciferase-templated formation of near-infrared emitting (800-1050 nm) PbS QDs [136]. This process, biomineralization, involves the incubation of QDs precursors with a solution of Luc8, which will serve as template for the self-assembly of the QD-Luc8 nanohybrid, without the need of other coupling methods. By BRET the blue light emitted by Luc8 is transferred to the so-formed QD and near-infrared light is emitted [136].

Initially the QD-Luc system was proposed for applications in bioimaging, but posteriorly it was suggested as bio-nanosensor for proteases [134, 137] and nucleic acids [138]. In the first case, Luc8 is coupled to the QDs through an animo acid sequence recognized by the protease. Without protease, the BRET signal will be detected; in its presence, the cleavage of the sequence will release Luc8 from QD and BRET ceases. In the second case, two oligonucleotide probes are assembled *in vitro*, one containing luciferase and the other a QD, being the QD-oligonucleotide sequence complementary to the target nucleic acid. Without nucleic acid, the two probes will hybridize, under certain conditions, and a BRET signal will be produced. In the presence of target nucleic acid, there will be a competition between the target and the Luc-probe to hybridize with QD-probe. The more target nucleic acid the more extensive will be its hybridization with QD-probe and lesser BRET signal will be produced.

All of these methods proved to be highly sensitive (limits of detection for the proteases of 1 ng mL^{-1} for MMP-2, 5 ng mL^{-1} for MMP-7 and 500 ng mL^{-1} for urokinase-type plasminogen activator [134], and 20 nM for nucleic acids [138]), simple and rapid (demanding one hour of incubation for protease detection and 30 minutes for nucleic acids assay) compared to other corresponding methods. They are also versatile, as other proteases and several oligonucleotides can be assayed just by changing the amino acid sequence between the luciferase and the QD or the nucleotidic sequence of the nucleic acid. Regarding bioimaging, it allowed the record of signals in a region of the electromagnetic spectrum where the light is less absorbed by tissues (red to near infrared), leading to improved signal-to-noise ratio and analysis at deeper locations inside living subjects, without performing any genetic alterations in luciferase. In fact, the QD-Luc8 self-illuminating system is so successful that are now commercially produced by Zymera, Inc, at San Jose, CA (USA) under the designation BRET-Qdot® [139].

Nanostructured Materials

Mesoporous Silica Nanoparticles

Silica, actually silicon dioxide, is one of the most common substances on Earth. It is present in rocks and soil and has important function in life, like in the diatoms. A broad range of applications involve silica, from basic pottery to photonic crystals (the self-assembly of spherical silica nanoparticles in a compact and periodic arrangement at nanometric scale that leads to diffraction of the incident light, resulting in different colours) [140-142].

In Nanoscience, silica is important in its colloidal state, in which silica particles (1-1000 nm) are dispersed in a continuous phase. If this phase is liquid, the system is called a sol. By aggregating these particles into a solid structure enclosing a continuous liquid phase the system is now called a gel [140]. These concepts are the basis of the sol-gel technology widely applied to prepare colloidal silica, for example the Stöber method, based on the basic hydrolysis of a silica precursor, leading to spherical and amorphous particles within 0.05-2 μm initially, but can be reduced up to 100-400 nm [140, 143].

Compared to other nanomaterials, for example carbon, silica is somehow limited in terms of functionalization, being the principal kind of functionalization related to organosilanes that will react with silanol groups (free hydroxyl groups derived from non-compensated, polymerized $Si(HO)_4$ at the surface) at silica's surface [140, 144, 145]. When using bifunctional organosilanes, one end will react with silanol and the other could have any functional group, so enhancing the applications. Furthermore, with a fine control of the synthesis process it can be attained the control of particle and pores' size and morphology, along with the insertion of functionalizations [140, 144, 145]. For example, porosity could be induced and controlled by using templates, generally surfactants, which are posteriorly removed [140, 144, 145]. Hybrid silica nanoparticles are also available, generally consisting in the coating of other nanoparticles with silica, for example magnetic hematite [146], or the production of nanocomposites, like copper-coated silica nanoparticles for odor removal [147] or gold-doped silica nanoparticles for biosensing [148].

Regarding luciferase, another method for ATP detection based on the immobilization of firefly luciferase into sugar-silica materials prepared form the sol-gel process was created [149]. The main improvement in this new method was that luciferase maintained a relatively high and stable light emission compared to other matrices like agarose beads and sepharose, a characteristic conferred by the covalent linkage of sugars (D-gluconolactone and D-maltonolactone) to the structure of the chosen silica precursor, (aminopropyl)triethoxysilane. The gel-containing luciferase can be re-used after several catalytical cycles, a unique feature compared to other systems, and it proved to be very sensitive, being able to detect 1 pM of ATP. In fact, the issue of enzyme stability upon immobilization needs attention, because this factor hampers the development of biosensors and bioreactors. A theoretical study analyzed the influence of temperature and composition of nanometric silica upon firefly luciferase's active site [150]. Using molecular dynamics simulation it was observed that nanoporous (6 nm width) hydrophilic silica can indeed help to stabilize luciferase at room (27 °C) and high (60 °C) temperatures.

Nanostructured Film

Another optical biosensor using nanospheres covered with a nanostructured film in which luciferase is embedded was proposed [151]. The film's assembly is based on the layer-by-layer method, which is based on the alternate adsorption of polycations and polyanions in solution to a desired template or substrate, which leads to the formation of multilayers whose composition and thickness can be finely controlled [152]. In this example [151], firefly luciferase and cationic poly(dimethyldiallyl ammonium chloride) were added to polystyrene sulfonated spheres with 520 nm of diameter and, through electrostatic interactions, several layers were deposited. Within 15 minutes, a monolayer was assembled. Results [151] showed that the immobilized luciferase retained about 70% of its activity compared to free enzymes. Furthermore, a sustained bioluminescence was detected during a 7-day analysis, especially if the outermost layer is composed of the cation, albeit with a reduced enzymatic activity in this particular case. Finally, these nanospheres were exposed to commercial solutions with different concentrations of ATP, and a response proportional to the ATP content was observed, suggesting a role as ATP sensor.

CONCLUSION

Nanoscience in general, and nanomaterials in particular, are the hot topics for the XXI century investigation. If it is true that a lot of work is already done, it is also true that many efforts are waiting to be solved out. For example, the interaction of nanoparticles with biomolecules within cells is poorly understood, although it is of paramount importance as more and more nanomaterials are being requested for biomedical applications [153, 154].

Although Bionanotechnology (or Nanobiotechnology) is taking its first steps it is already regarded as an area with a huge future ahead. In this context, the coupling of luciferases with nanomaterials can be regarded as a fruitful partnership. In fact, major advances have been observed in luciferase's enzymatic mechanism and biotechnology, which opened new areas for applications of the highly sensitive and selective luciferase-based bioanalytical methodologies. The main achievements so far were towards biosensing for ATP, nucleic acids and proteases, and bioimaging, but other applications are likely to arise. Carbon-based nanomaterials have the most promising potential for biological applications because of their biocompatibility and nontoxicity. Coupling these materials with well-established luciferase methodologies will undoubtedly be a good bet, not only for the optimization of already existing bioanalytical methodologies but also for the designing of new ones.

In conclusion nanomaterials, tiny entities which occupy "the region between the atomistic and the macroscopic worlds" [8], and luciferase, an indispensable bioanalytical tool, coupled to each other, represent the future of Bioanalytical Chemistry and Biomedicine.

REFERENCES

[1] Edwards, P. P. & Thomas, J. M. (2007). Gold in a metallic divided state-from Faraday to present-day nanoscience. *Angew. Chem.-Int. Edit.*, *46*, 5480-5486.

[2] The Royal Society & The Royal Academy of Engineering. Nanoscience and Nanotechnologies: opportunities and uncertainties. July 2004 [consulted at 10/13/2010]. Available from: http://www.nanotec.org.uk/finalReport.htm.

[3] Daniel, M.-C. & Astruc, D. (2004). Gold nanoparticles: assembly, supramolecular chemistry, quantum-size-related properties, and applications toward biology, catalysis, and nanotechnology. *Chem. Rev.*, *104*, 293-346.

[4] Valden, M., Lai, X. & Goodman, D. W. (1998). Onset of catalytic activity of gold clusters on titania with the appearance of nonmetallic properties. *Science*, *281*, 1647-1650.

[5] Brus, L. (1986). Electronic wave functions in semiconductor clusters: experiment and theory. *J. Phys. Chem.*, *90*, 2555-2560.

[6] Smith, A. M. & Nie, S. (2010). Semiconductor nanocrystals: structure, properties, and band gap engineering. *Accounts Chem. Res.*, *43*, 190-200.

[7] Gullapalli, H., Vemuru, V. S. M., Kumar, A., Botello-Mendez, A., Vajtai, R., Terrones, M., Nagarajaiah, S. & Ajayan, P. M. (2010). Flexible piezoelectric ZnO–paper nanocomposite strain sensor. *Small*, *6*, 1641-1646.

[8] Costi, R., Saunders, A. E. & Banin, U. (2010). Colloidal hybrid nanostructures: a new type of functional materials. *Angew. Chem.-Int. Edit.*, *49*, 4878-4897.

[9] Schulz-Drost, C., Sgobba, V., Gerhards, C., Leubner, S., Calderon, R. M. K., Ruland, A. & Guldi, D. M. (2010). Innovative inorganic–organic nanohybrid materials: coupling quantum dots to carbon nanotubes. *Angew. Chem.-Int. Edit.*, *49*, 6425-6429.

[10] Kang, H., Lim, S., Park, N., Chun, K.-Y. & Baik, S. (2010). Improving the sensitivity of carbon nanotube sensors by benzene functionalization. *Sens. Actuator B-Chem.*, *147*, 316-321.

[11] Ozin, G. A., Arsenault, A. C., & Cademartiri, L. (2009). *Nanochemistry - A Chemical Approach to Nanomaterials* (2nd edition). Cambridge, The Royal Society of Chemistry.

[12] Whitesides, G. M. & Boncheva, M. (2002). Beyond molecules: self-assembly of mesoscopic and macroscopic components. *Proc. Natl. Acad. Sci. U. S. A.*, *99*, 4769-4774.

[13] Whitesides, G. M. & Grzybowski, B. (2002). Self-assembly at all scales. *Science*, *295*, 2418-2421.

[14] Fialkowski, M., Bishop, K. J. M., Klajn, R., Smoukov, S. K., Campbell, C. J. & Grzybowski, B. A. (2006). Principles and implementations of dissipative (dynamic) self-assembly. *J. Phys. Chem. B*, *110*, 2482-2496.

[15] Roda, A., Guardigli, M., Michelini, E. & Mirasoli, M. (2009). Bioluminescence in analytical chemistry and *in vivo* imaging. *Trends Anal. Chem.*, *28*, 307-322.

[16] Hirokawa, K., Kajiyama, N. & Murakami, S. (2002). Improved practical usefulness of firefly luciferase by gene chimerization and random mutagenesis. *Biochim. Biophys. Acta-Protein Struct. Mol. Enzymol., 1597*, 271-279.

[17] Loening, A. M., Fenn, T. D., Wu, A. M. & Gambhir, S. S. (2006). Consensus guided mutagenesis of *Renilla* luciferase yields enhanced stability and light output. *Protein Eng. Des. Sel., 19*, 391-400.

[18] Fujii, H., Noda, K., Asami, Y., Kuroda, A., Sakata, M. & Tokida, A. (2007). Increase in bioluminescence intensity of firefly luciferase using genetic modification. *Anal. Biochem., 366*, 131-136.

[19] Welsh, J. P., Patel, K. G., Manthiram, K. & Swartz, J. R. (2009). Multiply mutated *Gaussia* luciferases provide prolonged and intense bioluminescence. *Biochem. Biophys. Res. Commun., 389*, 563-568.

[20] Caysa, H., Jacob, R., Müther, N., Branchini, B., Messerle, M. & Söling, A. (2009). A redshifted codon-optimized firefly luciferase is a sensitive reporter for bioluminescence imaging. *Photochem. Photobiol. Sci., 8*, 52-56.

[21] Branchini, B. R., Ablamsky, D. M., Davis, A. L., Southworth, T. L., Butler, B. Fan, F., Jathoul, A. P. & Pule, M. A. (2010). Red-emitting luciferases for bioluminescence reporter and imaging applications. *Anal. Biochem., 396*, 290–297.

[22] Riahi-Madvar, A. & Hosseinkhani, S. (2009). Design and characterization of novel trypsin-resistant firefly luciferases by site-directed mutagenesis. *Protein Eng. Des. Sel., 22*, 655-663.

[23] Imani, M., Hosseinkhani, S., Ahmadian, S. & Nazari, M. (2010). Design and introduction of a disulfide bridge in firefly luciferase: increase of thermostability and decrease of pH sensitivity. *Photochem. Photobiol. Sci., 9*, 1167-1177.

[24] Taneoka, A., Sakaguchi-Mikami, A., Yamazaki, T., Tsugawa, W. & Sode, K. (2009). The construction of a glucose-sensing luciferase. *Biosens. Bioelectron., 25*, 76-81.

[25] Zhang, Y., Phillips, G. J., Li, Q. & Yeung, E. S. (2008). Imaging localized astrocyte ATP release with firefly luciferase beads attached to the cell surface. *Anal. Chem., 80*, 9316-9325.

[26] Villalobos, V., Naik, S. & Piwnica-Worms, D. (2007). Current state of imaging protein-protein interactions *in vivo* with genetically encoded reporters. *Annu. Rev. Biomed. Eng., 9*, 321-49.

[27] Luker, K. E., Gupta, M. & Luker, G. D. (2008). Imaging CXCR4 signaling with firefly luciferase complementation. *Anal. Chem., 80*, 5565-5573.

[28] Kim, J., Moon, C. H., Jung, S. & Paik, S. R. (2009). α-Synuclein enhances bioluminescent activity of firefly luciferase by facilitating luciferin localization. *Biochim. Biophys. Acta – Proteins Proteomics, 1794*, 309-314.

[29] Fontes, R., Fernandes, D., Peralta, F., Fraga, H., Maio, I. & Esteves da Silva, J. C. G. (2008). Pyrophosphate and tripolyphosphate affect firefly luciferase luminescence because they act as substrates and not as allosteric effectors. *FEBS J., 275*, 1500-1509.

[30] Fraga, H., Fernandes, D., Fontes, R. & Esteves da Silva, J. C. G. (2005). Coenzyme A affects firefly luciferase luminescence because it acts as a substrate and not as an allosteric effector. *FEBS J., 272*, 5206-5216.

[31] Nakata, N., Ishida, A., Tani, H., Kamidate, T. (2003). Cationic liposomes enhanced firefly bioluminescent assay of bacterial ATP in the presence of an ATP extractant. *Anal. Sci., 19*, 1183-1185.

[32] Kheirolomoom, A., Kruse, D. E., Qin, S., Watson, K. E., Lai, C.-Y., Young, L. J. T., Cardiff, R. D. & Ferrara, K. W. (2010). Enhanced *in vivo* bioluminescence imaging using liposomal luciferin delivery system. *J. Control. Release,* 141, 128-136.
[33] Ganjalikhany, M. R., Ranjbar, B., Hosseinkhani, S., Khalifeh, K. & Hassani, L. (2010). Roles of trehalose and magnesium sulfate on structural and functional stability of firefly luciferase. *J. Mol. Catal. B-Enzym., 62,* 127-132.
[34] Ataei, F., Hosseinkhani, S. & Khajeh, K. (2009). Luciferase protection against proteolytic degradation: a key for improving signal in nano-system biology. *J. Biotechnol., 144,* 83-88.
[35] Shinde, R., Perkins, J. & Contag, C. H. (2006). Luciferin derivatives for enhanced *in vitro* and *in vivo* bioluminescence assays. *Biochemistry, 45,* 11103-11112.
[36] Shao, Q., Jiang, T., Ren, G., Cheng, Z. & Xing, B. (2009). Photoactivable bioluminescent probes for imaging luciferase activity. *Chem. Commun., 27,* 4028-4030.
[37] Gao, J. & Xu, B. (2009). Applications of nanomaterials inside cells. *Nano Today, 4,* 37-51.
[38] Sharma, P., Brown, S. C., Singh, A., Iwakuma, N., Pyrgiotakis, G., Krishna, V., Knapik, J. A., Barr, K., Moudgil, B. M. & Grobmyer, S. R. (2010). Near-infrared absorbing and luminescent gold speckled silica nanoparticles for photothermal therapy. *J. Mater. Chem., 20,* 5182-5185.
[39] Isobe, H., Nakanishi, W., Tomita, N., Jinno, S., Hiroto Okayama, H. & Nakamura, E. (2006). Nonviral gene delivery by tetraamino fullerene. *Mol. Pharm., 3,* 124-134.
[40] Frasco, M. F. & Chaniotakis, N. (2010). Bioconjugated quantum dots as fluorescent probes for bioanalytical applications. *Anal. Bioanal. Chem., 396,* 229-240.
[41] Krueger, A. (2010). *Carbon Materials and Nanotechnology.* Weinheim, Wiley-VHC.
[42] Loh, K. P., Bao, Q., Ang, P. K. & Yang, J. (2010). The chemistry of graphene. *J. Mater. Chem., 20,* 2277-2289.
[43] Soldano, C., Mahmood, A. & Dujardin, E. (2010). Production, properties and potential of graphene. *Carbon, 48,* 2127-2150.
[44] Yang, W., Ratinac, K. R., Ringer, S. P., Thordarson, P., Gooding, J. J. & Braet, F. (2010). Carbon nanomaterials in biosensors: should you use nanotubes or graphene?. *Angew. Chem.-Int. Edit., 49,* 2114-2138.
[45] Dresselhaus, M. S., Dresselhaus, G., & Avouris, P. Eds (2001). *Carbon Nanotubes - Synthesis, Structure, Properties, and Applications.* Leipzig, Springer.
[46] Santos, L. J., Rocha, G. P., Alves, R. B. & Freitas, R. P. (2010). Fulereno[C_{60}]: química e aplicações. Quim. Nova, 33, 680-693.
[47] Barnard, A. S. (2009). Diamond standard in diagnostics: nanodiamond biolabels make their mark. *Analyst,* 134, 1751-1764.
[48] Ryoo, R., Joo, S. H. & Jun, S. (1999). Synthesis of highly ordered carbon molecular sieves via template-mediated structural transformation. *J. Phys. Chem. B, 103,* 7743-7746.
[49] Ryoo, R., Joo, S. H., Kruk, M. & Jaroniec, M. (2001). Ordered mesoporous carbons. *Adv. Mater., 13,* 677-681.
[50] Stein, A., Wang, Z. & Fierke, M. A. (2009). Functionalization of porous carbon materials with designed pore architecture. *Adv. Mater., 21,* 265-293.
[51] Hirsch, A. (2002). Functionalization of single-walled carbon nanotubes. *Angew. Chem.-Int. Edit., 41,* 1853-1859.

[52] Balasubramanian, K. & Burghard, M. (2005). Chemically functionalized carbon nanotubes. *Small, 1*, 180-192.
[53] Tasis, D., Tagmatarchis, N., Bianco, A. & Prato, M. (2006). Chemistry of carbon nanotubes. *Chem. Rev., 106*, 1105-1136.
[54] Zhao, Y.-L. & Stoddart, J. F. (2009) Noncovalent functionalization of single-walled carbon nanotubes. *Accounts Chem. Res., 42*, 1161-1171.
[55] Serp, P., & Figueiredo, J. L. Eds (2009). *Carbon Materials for Catalysis*. Hoboken, John Wiley & Sons.
[56] Guglielmotti, V., Chieppa, S., Orlanducci, S., Tamburri, E., Toschi, F., Terranova, M. L. & Rossi, M. (2009). Carbon nanotube/nanodiamond structures: an innovative concept for stable and ready-to-start electron emitters. *Appl. Phys. Lett., 95*, 222113-222113-3.
[57] Lin, Y., Taylor, S., Li, H., Fernando, K. A. S., Qu, L., Wang, W., Gu, L., Zhou, B. & Sun, Y.-P. (2004). Advances toward bioapplications of carbon nanotubes. *J. Mater. Chem., 14*, 527-541.
[58] Katz, E. & Willner, I. (2004). Biomolecule-functionalized carbon nanotubes: applications in nanobioelectronics. *ChemPhysChem, 5*, 1084-1104.
[59] Cui, H.-F., Vashist, S. K., Al-Rubeaan, K., Luong, J. H. T. & Sheu, F.-S. (2010). Interfacing carbon nanotubes with living mammalian cells and cytotoxicity issues. *Chem. Res. Toxicol., 23*, 1131-1147.
[60] Tasis, D., Tagmatarchis, N., Georgakilas, V. & Prato, M. (2003). Soluble carbon nanotubes. *Chem.-Eur. J., 9*, 4000-4008.
[61] Kim, J.-H., Ahn, J.-H., Barone, P. W., Jin, H., Zhang, J., Heller, D. A. & Strano, M. S. (2010). A luciferase/single-walled carbon nanotube conjugate for near-infrared fluorescent detection of cellular ATP. *Angew. Chem.-Int. Edit., 49*, 1456-1459.
[62] Holt, K. B. (2010). Undoped diamond nanoparticles: origins of surface redox chemistry. *Phys. Chem. Chem. Phys., 12*, 2048-2058.
[63] Pichot, V., Stephan, O., Comet, M., Fousson, E., Mory, J., March, K. & Spitzer, D. (2010). High nitrogen doping of detonation nanodiamonds. *J. Phys. Chem. C, 114*, 10082-10087.
[64] Vlasov, I. I., Shenderova, O., Turner, S., Lebedev, O. I., Basov, A. A., Sildos, I., Rähn, M., Shiryaev, A. A. & Van Tendeloo, G. (2010). Nitrogen and luminescent nitrogen-vacancy defects in detonation nanodiamond. *Small, 6*, 687-694.
[65] http://www.alit.kiev.ua/index.htm. Consulted at 11/08/2010.
[66] Shimkunas, R. A., Robinson, E., Lam, R., Lu, S., Xu, X., Zhang, X.-Q., Huang, H., Osawa, E. & Ho, D. (2009). Nanodiamond-insulin complexes as pH-dependent protein delivery vehicles. *Biomaterials, 30*, 5720-5728.
[67] Li, J., Zhu, Y., Li, W., Zhang, X., Peng, Y. & Huang, Q. (2010). Nanodiamonds as intracellular transporters of chemotherapeutic drug. *Biomaterials, 31*, 8410-8418.
[68] Zhang, Q., Mochalin, V. N., Neitzel, I., Knoke, I. Y., Han, J., Klug, C. A., Zhou, J. G., Lelkes, P. I. & Gogotsi, Y. (2011). Fluorescent PLLA-nanodiamond composites for bone tissue engineering. *Biomaterials, 32*, 87-94.
[69] Fu, C.-C., Lee, H.-Y., Chen, K., Lim, T.-S., Wu, H.-Y., Lin, P.-K., Wei, P.-K., Tsao, P.-H., Chang, H.-C. & Fann, W. (2007). Characterization and application of single fluorescent nanodiamonds as cellular biomarkers. *Proc. Natl. Acad. Sci. U. S. A., 104*, 727-732.

[70] Zhao, W., Xu, J.-J., Qiu, Q.-Q. & Chen, H.-Y. (2006). Nanocrystalline diamond modified gold electrode for glucose biosensing. *Biosens. Bioelectron.*, *22*, 649-655.

[71] Raina, S., Kang, W. P. & Davidson, J. L. (2010). Fabrication of nitrogen-incorporated nanodiamond ultra-microelectrode array for dopamine detection. *Diam. Relat. Mat.*, *19*, 256-259.

[72] Puzyr', A. P., Pozdnyakova, I. O. & Bondar', V. S. (2004). Design of a luminescent biochip with nanodiamonds and bacterial luciferase. *Phys. Solid State*, *46*, 761-763.

[73] Feldheim, D. L., & Foss Jr., C. A. Eds (2002). *Metal Nanoparticles – Synthesis, Characterization, and Applications*. New York, Marcel Dekker.

[74] Lisiecki I. (2005). Size, shape, and structural control of metallic nanocrystals. *J. Phys. Chem. B*, *109*, 12231-12244.

[75] Grzelczak, M., Pérez-Juste, J., Mulvaney, P. & Liz-Marzán, L. M. (2008). Shape control in gold nanoparticle synthesis. *Chem. Soc. Rev.*, *37*, 1783-1791.

[76] Meng, X. K., Tang, S. C. & Vongehr, S. (2010). A review on diverse silver nanostructures. *J. Mater. Sci. Technol.*, *26*, 487-522.

[77] Kulkarni, A. P., Noone, K. M., Munechika, K., Guyer, S. R. & Ginger, D. S. (2010). Plasmon-enhanced charge carrier generation in organic photovoltaic films using silver nanoprisms. *Nano Lett.*, *10*, 1501-1505.

[78] Chow, P. E. Ed (2010). *Gold Nanoparticles - Properties, Characterization and Fabrication*. Hauppauge, Nova Science Publishers.

[79] Panyala, N. R., Peña-Méndez, E. M. & Havel, J. (2009). Gold and nano-gold in medicine: overview, toxicology and perspectives. *J. Appl. Biomed.*, *7*, 75-91.

[80] Giljohann, D. A., Seferos, D. S., Daniel, W. L., Massich, M. D., Patel, P. C. & Mirkin, C. A. (2010). Gold nanoparticles for biology and medicine. *Angew. Chem.-Int. Edit.*, *49*, 3280-3294.

[81] Goesmann, H. & Feldmann, C. (2010). Nanoparticulate functional materials. *Angew. Chem.-Int. Edit.*, *49*, 1362-1395.

[82] Love, J. C., Estroff, L. A., Kriebel, J. K., Nuzzo, R. G. & Whitesides, G. M. (2005). Self-assembled monolayers of thiolates on metals as a form of nanotechnology. *Chem. Rev.*, *105*, 1103-1169.

[83] Tokonami, S., Morita, N., Takasaki, K. & Toshima, N. (2010). Novel synthesis, structure, and oxidation catalysis of Ag/Au bimetallic nanoparticles. *J. Phys. Chem. C*, *114*, 10336-10341.

[84] Zhang, S., Shao, Y., Yin, G. & Lin, Y. (2010). Electrostatic self-assembly of a Pt-around-Au nanocomposite with high activity towards formic acid oxidation. *Angew. Chem.-Int. Edit.*, *49*, 2211-2214.

[85] Huang, C., Jiang, J., Muangphat, C., Sun, X. & Hao, Y. (2011). Trapping iron oxide into hollow gold nanoparticles. *Nanoscale Res. Lett.*, *6*, 1-5.

[86] Aslan, K., Lakowicz, J. R. & Geddes, C. D. (2005). Plasmon light scattering in biology and medicine: new sensing approaches, visions and perspectives. *Curr. Opin. Chem. Biol.*, *9*, 538-544.

[87] Lee, S. E., Liu, G. L., Kim, F. & Lee, L. P. (2009). Remote optical switch for localized and selective control of gene interference. *Nano Lett.*, *9*, 562-570.

[88] Gobin, A. M., Watkins, E. M., Quevedo, E., Colvin, V. L. & West, J. L. (2010). Near-infrared-resonant gold/gold sulfide nanoparticles as a photothermal cancer therapeutic agent. *Small*, *6*, 745-752.

[89] Wang, S., Singh, A. K., Senapati, D., Neely, A., Yu, H. & Ray, P. C. (2010). Rapid colorimetric identification and targeted photothermal lysis of *Salmonella* bacteria by using bioconjugated oval-shaped gold nanoparticles. *Chem.-Eur. J.*, *16*, 5600-5606.

[90] Leong, W. L., Lee, P. S., Mhaisalkar, S. G., Chen, T. P, & Dodabalapur, A. (2007). Charging phenomena in pentacene-gold nanoparticle memory device. *Appl. Phys. Lett.*, *90*, 042906-042906-3.

[91] Cortez, J., Vorobieva, E., Gralheira, D., Osório, I., Soares, L., Vale, N., Pereira, E., Gomes, P. & Franco, R. Bionanoconjugates of tyrosinase and peptide-derivatised gold nanoparticles for biosensing of phenolic compounds. *J. Nanopart. Res.*, DOI: 10.1007/s11051-010-0099-8.

[92] Kim, Y.-P., Daniel, W. L., Xia, Z., Xie, H., Mirkin, C. A. & Rao, J. (2010). Bioluminescent nanosensors for protease detection based upon gold nanoparticle-luciferase conjugates. *Chem. Commun.* *46*, 76-78.

[93] Fingleton, B. (2006). Matrix metalloproteinases: roles in cancer and metastasis. *Front. Biosci.*, *11*, 479-491.

[94] Bimberg, D., Grundmann, M., & Ledentsov, N. N. (1999). *Quantum Dot Heterostructures*. Salisbury, John Wiley & Sons.

[95] Sun, Y.-P., Zhou, B., Lin, Y., Wang, W., Fernando, K. A. S., Pathak, P., Meziani, M. J., Harruff, B. A., Wang, X., Wang, H., Luo, P. G., Yang, H., Kose, M. E., Chen, B., Veca, L. M. & Xie, S.-Y. (2006). Quantum-sized carbon dots for bright and colorful photoluminescence. *J. Am. Chem. Soc.*, *128*, 7756-7757.

[96] Baker, S. N. & Baker, G. A. (2010). Luminescent carbon nanodots: emergent nanolights. *Angew. Chem.-Int. Edit.*, *49*, 6721-6744.

[97] Yu-Juan, J., Yun-Jun, L., Guo-Ping, L., Jie, L., Yuan-Feng, W., Rui-Qin, Y. & Wen-Ting, L. (2008). Application of photoluminescent CdS/PAMAM nanocomposites in fingerprint detection. *Forensic Sci. Int.*, *179*, 34-38.

[98] Dilag, J., Kobus, H. & Ellis, A. V. (2009). Cadmium sulfide quantum dot/chitosan nanocomposites for latent fingermark detection. *Forensic Sci. Int.*, *187*, 97-102.

[99] Hu, X., Han, H., Hua, L. & Sheng, Z. (2010). Electrogenerated chemiluminescence of blue emitting ZnSe quantum dots and its biosensing for hydrogen peroxide. *Biosens. Bioelectron.*, *25*, 1843-1846.

[100] Nabiev, I., Rakovich, A., Sukhanova, A., Lukashev, E., Zagidullin, V., Pachenko, V., Rakovich, Y. P., Donegan, J. F., Rubin, A. B. & Govorov, A. O. (2010). Fluorescent quantum dots as artificial antennas for enhanced light harvesting and energy transfer to photosynthetic reaction centers. *Angew. Chem.-Int. Edit.*, *49*, 7217-7221.

[101] Bruchez Jr., M., Moronne, M., Gin, P., Weiss, S. & Alivisatos, A. P. (1998). Semiconductor nanocrystals as fluorescent biological labels. *Science*, *281*, 2013-2016.

[102] Smith, A. M., Duan, H., Mohs, A. M. & Nie, S. (2008). Bioconjugated quantum dots for *in vivo* molecular and cellular imaging. *Adv. Drug Deliv. Rev.*, *60*, 1226-1240.

[103] Juzenas, P., Chen, W., Sun, Y.-P., Coelho, M. A. N., Generalov, R., Generalova, N. & Christensen, I. L. (2008). Quantum dots and nanoparticles for photodynamic and radiation therapies of cancer. *Adv. Drug Deliv. Rev.*, *60*, 1600-1614.

[104] Algar, W. R., Tavares, A. J. & Krull, U. J. (2010). Beyond labels: a review of the application of quantum dots as integrated components of assays, bioprobes, and biosensors utilizing optical transduction. *Anal. Chim. Acta*, *673*, 1-25.

[105] Wagner, M. K., Li, F., Li, J., Li, X.-F. & Le, X. C. (2010). Use of quantum dots in the development of assays for cancer biomarkers. *Anal. Bioanal. Chem.*, *397*, 3213-3224.

[106] Tholouli, E., Sweeney, E., Barrow, E., Clay, V., Hoyland, J. A. & Byers, J. R. (2008). Quantum dots light up pathology. *J. Pathol.*, *216*, 275-285.

[107] Liu, W., Howarth, M., Greytak, A. B., Zheng, Y., Nocera, D. G., Ting, A. Y. & Bawendi, M. G. (2008). Compact biocompatible quantum dots functionalized for cellular imaging. *J. Am. Chem. Soc.*, *130*, 1274-1284.

[108] Gao, J., Chen, K., Xie, R., Xie, J., Lee, S., Cheng, Z., Peng, X. & Chen, X. (2010). Ultrasmall near-infrared non-cadmium quantum dots for *in vivo* tumor imaging. *Small*, *6*, 256-261.

[109] Crane, J. M., Van Hoek, A. N., Skach, W. R. & Verkman, A. S. (2008). Aquaporin-4 dynamics in orthogonal arrays in live cells visualized by quantum dot single particle tracking. *Mol. Biol. Cell*, *19*, 3369-3378.

[110] You, C., Wilmes, S., Beutel, O., Löchte, S., Podoplelowa, Y., Roder, F., Richter, C., Seine, T., Schaible, D., Uzé, G., Clarke, S., Pinaud, F., Dahan, M. & and Piehler, J. (2010). Self-controlled monofunctionalization of quantum dots for multiplexed protein tracking in live cells. *Angew. Chem.-Int. Edit.*, *49*, 4108-4112.

[111] Luo, K., Li, S., Xie, M., Wu, D., Wang, W., Chen, R., Huang, L., Huang, T., Pang, D. & Xiao, G. (2010). Real-time visualization of prion transport in single live cells using quantum dots. *Biochem. Biophys. Res. Commun.*, *394*, 493-497.

[112] Freeman, R., Finder, T., Gill, R. & Willner, I. (2010). Probing protein kinase (CK2) and alkaline phosphatase with CdSe/ZnS quantum dots. *Nano Lett.*, *10*, 2192-2196.

[113] Wu, P., He, Y., Wang, H.-F. & Yan, X.-P. (2010). Conjugation of glucose oxidase onto Mn-doped ZnS quantum dots for phosphorescent sensing of glucose in biological fluids. *Anal. Chem.*, *82*, 1427-1433.

[114] Derfus, A. M., Chan, W. C. W. & Bhatia, S. N. (2004). Probing the cytotoxicity of semiconductor quantum dots. *Nano Lett.*, *4*, 11-18.

[115] Hoshino, A., Fujioka, K., Oku, T., Suga, M., Sasaki, Y. F., Ohta, T., Yasuhara, M., Suzuki, K. & Yamamoto, K. (2004). Physicochemical properties and cellular toxicity of nanocrystal quantum dots depend on their surface modification. *Nano Lett.*, *4*, 2163-2169.

[116] Lovrić, J., Cho, S. J., Winnik, F. M. & Maysinger, D. (2005). Unmodified cadmium telluride quantum dots induce reactive oxygen species formation leading to multiple organelle damage and cell death. *Chem. Biol.*, *12*, 1227-1234.

[117] Lovrić, J., Bazzi, H. S., Cuie, Y., Fortin, G. R. A., Winnik, F. M. & Maysinger, D. (2005). Differences in subcellular distribution and toxicity of green and red emitting CdTe quantum dots. *J. Mol. Med.*, *83*, 377-385.

[118] Cho, S. J., Maysinger, D., Jain, M., Röder, B., Hackbarth, S. & Winnik, F. M. (2007). Long-term exposure to CdTe quantum dots causes functional impairments in live cells. *Langmuir*, *23*, 1974-1980.

[119] Susumu, K., Uyeda, H. T., Medintz, I. L., Pons, T., Delehanty, J. B. & Mattoussi, H. (2007). Enhancing the stability and biological functionalities of quantum dots via compact multifunctional ligands. *J. Am. Chem. Soc.*, *129*, 13987-13996.

[120] Pradhan, N., Goorskey, D., Thessing, J., & Peng, X. (2005). An alternative of CdSe nanocrystal emitters: pure and tunable impurity emissions in ZnSe nanocrystals. *J. Am. Chem. Soc.*, *127*, 17586-17587.

[121] Cai, W., Hsu, A. R., Li, Z.-B. & Chen, X. (2007). Are quantum dots ready for in vivo imaging in human subjects?. *Nanoscale Res. Lett.*, *2*, 265-281.

[122] Hale, G. M. & Querry, M. R. (1973). Optical constants of water in the 200-nm to 200-μm wavelength region. *Appl. Optics*, *12*, 555-563.

[123] So, M.-K., Xu, C., Loening, A. M., Gambhir, S. S. & Rao, J. (2006). Self-illuminating quantum dot conjugates for *in vivo* imaging. *Nat. Biotechnol.*, *24*, 339-343.

[124] Förster, T. (1948). Zwischenmolekulare energiewanderung und fluoreszenz. *Ann. Phys. – Berlin*, *437*, 55-75.

[125] Roda, A., Guardigli, M., Michelini, E. & Mirasoli, M. (2009). Nanobioanalytical luminescence: Förster-type energy transfer methods. *Anal. Bioanal. Chem.*, *393*, 109-123.

[126] Medintz, I. L. & Mattoussi, H. (2009). Quantum dot-based resonance energy transfer and its growing application in biology. *Phys. Chem. Chem. Phys.*, *11*, 17-45.

[127] Xu, Y., Piston, D. W. & Johnson, C. H. (1999). A bioluminescence resonance energy transfer (BRET) system: application to interacting circadian clock proteins. *Proc. Natl. Acad. Sci. U. S. A.*, *96*, 151-156.

[128] Bacart, J., Corbel, C., Jockers, R., Bach, S. & Couturier, C. (2008). The BRET technology and its application to screening assays. *Biotechnol. J.*, *3*, 311-324.

[129] Xia, Z. & Rao, J. (2009). Biosensing and imaging based on bioluminescence resonance energy transfer. *Curr. Opin. Biotechnol.*, *20*, 37-44.

[130] Chen, L., Algar, W. R., Tavares, A. J. & Krull, U. J. (2011). Toward a solid-phase nucleic acid hybridization assay within microfluidic channels using immobilized quantum dots as donors in fluorescence resonance energy transfer. *Anal. Bioanal. Chem.*, *399*, 133-141.

[131] Hering, V. R., Gibson, G., Schumacher, R. I., Faljoni-Alario, A. & Politi, M. J. (2007). Energy transfer between CdSe/ZnS core/shell quantum dots and fluorescent proteins. *Bioconjugate Chem.*, *18*, 1705-1708.

[132] Dong, H., Gao, W., Yan, F., Ji, H. & Ju, H. (2010). Fluorescence resonance energy transfer between quantum dots and graphene oxide for sensing biomolecules. *Anal. Chem.*, *82*, 5511-5517.

[133] Zhang, Y., So, M.-K., Loening, A. M., Yao, H., Gambhir, S. S. & Rao, J. (2006). HaloTag protein-mediated site-specific conjugation of bioluminescent proteins to quantum dots. *Angew. Chem.-Int. Edit.*, *45*, 4936-4940.

[134] Xia, Z., Xing, Y., So, M.-K., Koh, A. L., Sinclair, R. & Rao, J. (2008). Multiplex detection of protease activity with quantum dot nanosensors prepared by intein-mediated specific bioconjugation. *Anal. Chem.*, *80*, 8649-8655.

[135] Xing, Y., So, M.-K., Koh, A. L., Sinclair, R. & Rao, J. (2008). Improved QD-BRET conjugates for detection and imaging. *Biochem. Biophys. Res. Commun.*, *372*, 388-394.

[136] Ma, N., Marshall, A. F. & Rao, J. (2010). Near-infrared light emitting luciferase *via* biomineralization. *J. Am. Chem. Soc.*, *132*, 6884-6885.

[137] Yao, H., Zhang, Y., Xiao, F., Xia, Z. & Rao, J. (2007). Quantum dot/bioluminescence resonance energy transfer based highly sensitive detection of proteases. *Angew. Chem.-Int. Edit.*, *46*, 4346-4349.

[138] Cissell, K. A., Campbell, S. & Deo, S. K. (2008). Rapid, single-step nucleic acid detection. *Anal. Bioanal. Chem.*, *391*, 2577-2581.

[139] http://www.zymera.com/. Consulted at 11/09/2010.

[140] Bergna, H. E. & Roberts, W. O. Eds (2006). *Colloidal Silica - Fundamentals and Applications*. Boca Raton, Taylor & Francis.

[141] Norris, D. J., Arlinghaus, E. G., Meng, L., Heiny, R. & Scriven, L. E. (2004). Opaline photonic crystals: how does self-assembly work?. *Adv. Mater.*, *16*, 1393-1399.

[142] Bardosova, M. & R. H. Tredgold, R. H. (2002). Ordered layers of monodispersive colloids. *J. Mater. Chem.*, *12*, 2835-2842.

[143] Stöber, W., Fink, A. & Bohn, E. (1968). Controlled growth of monodisperse silica spheres in the micron size range. *J. Colloid Interface Sci.*, *26*, 62-69.

[144] Fryxell, G. E. (2006). The synthesis of functional mesoporous materials. *Inorg. Chem. Commun.*, *9*, 1141-1150.

[145] Slowing, I. I., Vivero-Escoto, J. L., Trewny, B. G. & Lin, V. S.-Y. (2010). Mesoporous silica nanoparticles: structural design and applications. *J. Mater. Chem.*, *20*, 7924-7937.

[146] Zhang, J., Thurber, A., Hanna, C. & Punnoose, A. (2010). Highly shape-selective synthesis, silica coating, self-assembly, and magnetic hydrogen sensing of hematite nanoparticles. *Langmuir*, *26*, 5273-5278.

[147] Singh, A., Krishna, V., Angerhofer, A., Do, B., MacDonald, G. & Moudgil, B. (2010). Copper coated silica nanoparticles for odor removal. *Langmuir*, *26*, 15837-15844.

[148] Lee, K. G., Wi, R., Park, T. J., Yoon, S. H., Lee, J., Lee, S. J. & Kim, D. H. (2010). Synthesis and characterization of gold-deposited red, green and blue fluorescent silica nanoparticles for biosensor application. *Chem. Commun.*, *46*, 6374-6376.

[149] Cruz-Aguado, J. A., Chen, Y., Zhang, Z., Elowe, N. H., Brook, M. A. & Brennan, J. D. (2004). Ultrasensitive ATP detection using firefly luciferase entrapped in sugar-modified sol-gel-derived silica. *J. Am. Chem. Soc.*, *126*, 6878-6879.

[150] Nishiyama, K. (2008). Thermal behavior of luciferase on nanofabricated hydrophilic Si surface. *Biomacromolecules*, *9*, 1081-1083.

[151] Pastorino, L., Disawal, S., Nicolini, C., Lvov, Y. M. & Erokhin, V. V. (2003). Complex catalytic colloids on the basis of firefly luciferase as optical nanosensor platform. *Biotechnol. Bioeng. 84*, 286-291.

[152] Siqueira Jr., J. R., Caseli, L., Crespilho, F. N., Zucolotto, V. & Oliveira Jr, O. N. (2010). Immobilization of biomolecules on nanostructured films for biosensing. *Biosens. Bioelectron.*, *25*, 1254-1263.

[153] Lynch, I., Cedervall, T., Lundqvist, M., Cabaleiro-Lago, C., Linse, S. & Dawson, K. A. (2007). The nanoparticles-protein complex as a biological entity; a complex fluids and surface science challenge for the 21st century. *Adv. Colloid Interface Sci.*, *134-135*, 167-174.

[154] Mailänder, V. & Landfester, K. (2009). Interaction of nanoparticles with cells. *Biomacromolecules*, *10*, 2379-2400.

In: Bioluminescence
Editor: David J. Rodgerson, pp. 49-70
ISBN 978-1-61209-747-3

Chapter 3

BIOLUMINESCENCE BIOREPORTER STRAINS FOR THE DETECTION OF QUORUM-SENSING *N*-ACYL-HOMOSERINE LACTONE SIGNAL MOLECULES

Michael A. Savka **and** ***André O. Hudson***
School of Biological and Medical Sciences,
Rochester Institute of Technology,
Rochester, NY 14623, USA

ABSTRACT

Recombinant whole-cell bioluminescence biosensors developed over the last decade have been very useful to discover bacteria that posses an N-acyl-homoserine lactone (acyl-HSL) regulated quorum-sensing (QS) communication system.The acyl-HSL-dependent bacterial bioreporter systems, also known as specific biosensors were constructed to respond to the presence of a certain small molecules, in this case acyl-HSLs.These acyl-HSL-dependent specific biosensors have been developed becauseacyl-HSL signalsdo not contain strongly absorbing chromophores and acyl-HSLs are difficult to detect with standard chemical tests.In addition, the problem of acyl-HSL detection ischallenging since these signals are produced at very low levels, customarily in the nanomolar to micromolar range. Acyl-HSLbiosensors contain an acyl-HSL signal-dependent receptor / regulator protein that is a member of the LuxR-family and a gene coding for a detectable response phenotype, the reporter gene. The expression of the reporter phenotype occurs only in the presence of acyl-HSL signals and is controlled by anacyl-HSL-LuxR complex responsive promoter fused to the reporter gene(s).A number of bioreporter strains, each containing a different LuxR homolog receptor, have been constructed that report light production or bioluminescence from the *luxCDABE* operon as detection end points.This review aims to provide a background to examine the characteristics and adaptations of bioluminescent bacterialbiosensors for acyl-HSL signal detection, quantification and tentative identification of acyl-HSLs.

* Corresponding author:Michael A. Savka; Telephone: 585-475-5141; FAX: 585-475-5760; email: massbi@rit.edu

Keywords: Autoinducer, bacterial communication, bioluminescence reporter strains, cell-to-cell communication,*luxCDABE*operon, LuxR, acyl-HSL-dependent receptors, signal molecules

ABBREVIATIONS

acyl-HSL, *N*-acyl-homoserine lactones; QS, quorum-sensing

1. INTRODUCTION

Whole-cell bacterial biosensors (bioreporters) respond to produce a quantifiable output in response to certain target chemicals.A 'bacterial bioreporter' has become to mean a genetically recombinant bacterium in which one or more regulatory network is used to produce a recognizable and quantifiable output or phenotype.This enables the native signaling pathway to be monitored by an artificial output.For this cellular process to occur, a gene or group of genes encode(s) the artificial output or detectable phenotype such as light production or bioluminescence.This output is coupled to the ability of the sensory-regulatory systems or receptor / regulator network to interact with the specific compound(s) being studied.The amount or activity of the detectable phenotype is accepted to be a measure of the cellular response to the target.Ideal outputs are easy to detect, highly sensitive, nontoxic, quantifiable and not present in the native bacterium that constitutes the bioreporter system.

On the basis of detection specificity, bioreporters have been divided into three groups: 1. Non-specific,2. Semi-specific and 3. Specific.

1.1. Non-specific

Non-specific bioreporters contain a reporter gene fused to a constitutive promoter and the toxicity of the sample is measured by the decrease in reporter protein output.Non-specific bioreporters are useful for measuring the general toxicity of a sample and unpredictable additive or synergistic effects between chemicals in complex mixtures and environmental samples.However, they cannot provide information about the identity of the contaminants in the sample (Dalzell *et al.*, 2002; Preston *et al.*, 2000; Unge *et al.*, 1999; Ulitzur *et al.*, 2002).

1.2. Semi-specific

Semi-specific bioreporters are designed to enable a toxic agent to induce a general stress response in the cell.The presence of contaminants in the sample is measured by an increase in reporter output production form a stress-induced promoter.Examples of these include: the heat shock and oxidative stress promoters.Membrane damage promoters such as the SOS- and the *ada*-controlled promoters are also included in this category (Sagi *et al.*, 2003; Mitchell and Gu, 2004; Norman *et al.*, 2005; Lee and Gu, 2003).

1.3. Specific

Specific bioreportersenable the perception of an effector chemical upon the cell into a specific expression of a reporter or detectable phenotype.The specificity of the bioreporter is achieved by coupling effector compound sensing with an activator or repressor receptor / regulator which is able to detect and measure a phenotype (Daunert *et al.,* 2000; van der Meek *et al.,* 2004; van der Meer and Belkin 2010).

Acyl-HSL signals are used in a gene regulatory process called quorum-sensing (QS) in many Gram-negative bacteria that are important in human, plant, animal and environmental processes (Atkinson and Williams, 2009).In this review we present an overview of specific bioluminescence whole-cell bacterial bioreporters for the detection, quantification and putative identification of bacterial cell-to-cell communication signals of Gram-negative bacteria known as *N*-acyl-homoserine lactones (acyl-HSLs).

1.4. Whole-cell Bioreporters Based on the *luxCDABE* Bioluminescence Output

Whole-cell specific bioreporters using bioluminescence as the reporter system was first developed to measure and report the presence of aromatic hydrocarbon contamination and naphthalene degradation from industrial soils twenty years ago (King *et al.,* 1990).This was the first bioreporter that enabled specific and single compound detection. The reporter system then used the recently cloned*Vibrio fischeri*bioluminescence operon named *lux*. The approach was considered innovative at that time.This was because the system was noninvasive, nondestructive, rapid and specific.This specificity relied on the insertion of the *lux* operon as a transcriptional fusion to the *nahG* gene encoding salicylate hydroxylase from *Pseudomonas fluorescens* (King *et al.,* 1990).

Bacterial luciferase encoded by a*luxCDABE* operon enables the oxidation of long-chain fatty aldehydes and reduced flavin mononucleotides (FMN) to form the corresponding fatty acid and FMN in the presence of molecular oxygen.This reaction results in the emission of a blue-green light with a maximum intensity at 490 nm (Fletcher *et al.,* 2007) and quantum efficiency between 0.05 and 0.15.The *luxCDABE* operon contains five genes, *luxC, D, A, B,* and *E*.The catalytic subunits, α and β, are encoded by the *luxA* and *luxB* genes.The expression of *luxC, D,* and *E* are necessary to form the fatty acid reductases that are required for the synthesis and recycling of the fatty aldehydes (Meighen, 1991;Wilson and Hastings 1998).The expression of the *luxCDABE* operon confers a bioluminescent or light producing phenotype on living bacterial cells that is highly sensitive without the addition of an external substrate.

Since the time of the first specific bioreporter, the scientific community has revealed a much better understanding of the different levels of regulatory control in bacteria, of the natural variety of sensory and regulatory proteins and of the mechanism of signal transduction that link specific compound perception to specific gene expression, a biological network.Furthermore the development of computational approaches to receptor design to manipulate molecular recognition between ligands and proteins is materializing in the advancement of biotechnological applications in the construction of novel and specific enzymes, genetic circuits, signal transduction pathways and biosensors.This advanced breadth of knowledge in bacteriology now provides a richer box of tools for the scientific development of ultra-sensitive and specific bacterial bioreporters (Looger *et al.,* 2003).

2. Cell-to-cell Communicationin Gram-negative Bacteria

Many bacteria employ a cell-to-cell communication system that relies on a cell density sensing mechanism called quorum sensing (QS) (Smith *et al.*, 2006; Taga and Bassler, 2003; Williams, 2007).Gram-negative bacteria have been shown to produce and respond to *N*-acyl-homoserine lactone (acyl-HSL) signal molecules that are associated with gene expression in QS regulation (Smith *et al.,* 2006).Acyl-HSL-specific QS in Gram-negative bacteria implicate two proteins, an acyl-HSL synthase (LuxI ortholog), and an acyl-HSL-responsive transcription factor (LuxR ortholog) that are involved with regulation of target genes.LuxR orthologs contain acyl-HSL-specific binding domains.*luxI* and*luxR* orthologues are often identified at the same genetic locus (Case *et al.,* 2008).However, more recently many *luxR* solos (*luxR* regulator in the absence of *luxI*acyl-HSL synthase) have been documented (Fuqua, 2006; Lee *et al.,*2006;Lequette *et al.*, 2006; Duerkop *et al.,* 2007; Ferluga *et al.*, 2007; Subramoni and Venturi 2009).As the population density of a bacterial community increases, the acyl-HSLs accumulate to physiologically relevant threshold concentrations that allow interaction with cognate transcription factors (Williams, 2007).In most cases, signal-bound LuxR-type proteins activate the target genes (Stevens and Greenberg 1997) while others function as transcriptional repressors in a ligand-free state (Minogue *et al.,* 2002).The overall result is the coordinated activation or repression of specific target genes in response to a bacterial quorum.QS systems are diverse and regulate a range of different virulence functions that affect human, plant and animal diseases (Cataldi *et al.,* 2007; Loh *et al.,* 2002; Ramey *et al.,* 2004; Asad and Opal, 2010).

2.1. Gram-negative Bacteria*n*-acyl-homoserine Lactone(acyl-HSL)Signals

The LuxI-type family of proteins are the major class of enzymes responsible for acyl-HSL biosynthesis (Ortori *et al.,* 2006).The bacterial database currently has over 100 LuxI members.Theseacyl-HSL synthase directs the synthesis of acyl-HSLs by catalyzing an amide bond between the appropriately charged acyl-acyl carrier proteins, the source of the fatty acyl side chain and*S*-adenosylmethionine (SAM), the source of the homoserine lactone ring (More *et al.,* 1996, Jiang *et al.,* 1998, Parsek and Greenberg 1999).Acyl-HSL biosynthesis directed by LuxI-type proteins is shown in Figure 1.

The two other acyl-HSL synthase familiesinclude AinS from *Vibrio fischeri*, VanM from *V. anguillarum* and LuxM from *V. harveyi* (Bassler *et al.,* 1994; Milton *et al.,* 2001) and the HdtS protein from *Pseudomonas fluorescens* (Laue *et al.,* 2000).The Vibrio family can use acyl-CoA or acyl-ACP and SAM as substrates (Hanzelka *et al.,* 1999).The third family represented by the HdtS protein is related to the lysophosphatidic acid acyltransferases family (Laue *et al.,* 2000).

Acyl-HSL signals are characterized by a homoserine lactone moiety ligated to an acyl side chain.The specificity determinants of the acyl-HSL signals are the length of the acyl side chain ranging from 4 to 18 carbons, substitution at the 3rd carbon position and saturation levels within the acyl chain (Fuqua and Winans, 1999; Tage and Bassler, 2003).Acyl-HSLs are characterized as long-chain or short-chain acyl-HSLs depending on whether their acyl moiety consists of greater than, or equal or less than eight carbons, respectively (Scott *et al.*, 2006).The structures, names and abbreviations of some acyl-HSLs are presented in Figure 2.

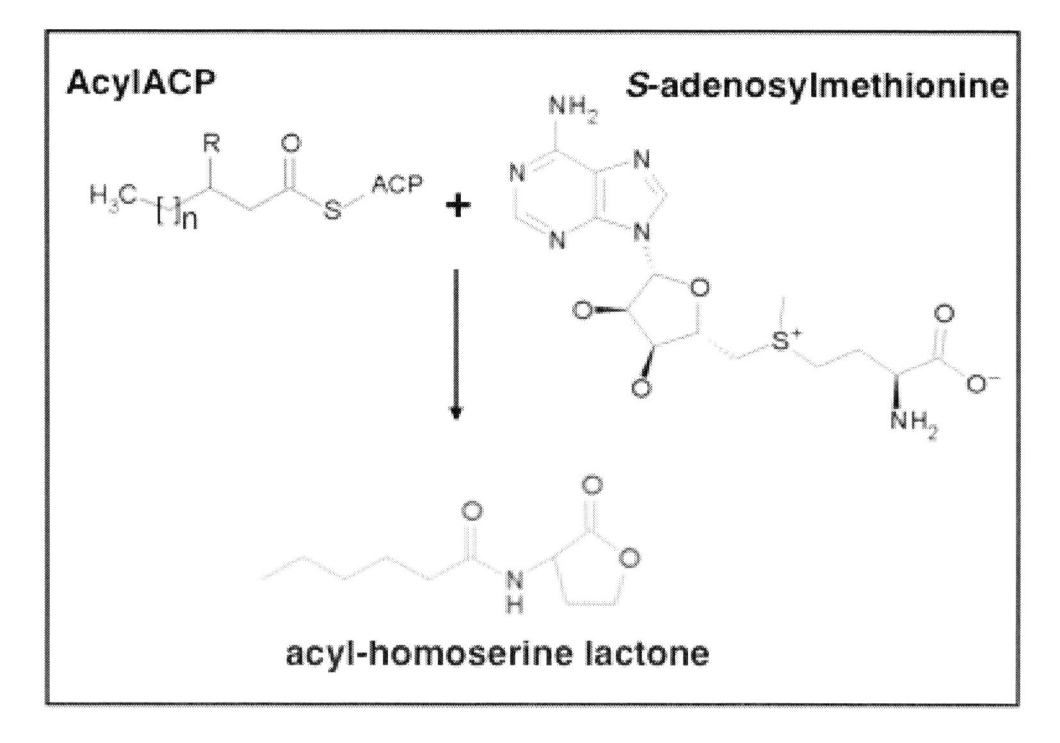

Figure 1.Biosynthesis of *N*-acyl-homoserine lactones (acyl-HSLs).Precursors of acylated acyl carrier protein and *S*-adenosylmethionine are substrates for LuxI homologues that produce acyl-HSLs.The "R" indicates the position of the third carbon in the acyl side chain of the acyl-HSL and the substitution site that can be a hydroxyl, a carbonyl group for fully reduced in different bacteria.The "n" indicates the number of carbons, which can vary in the acyl side chain between 4 to 18 carbons.An additional modification can include unsaturated bonds in the acyl side chain.

2.2. LuxR-type acyl-HSL-specific Receptor Proteins

All of the regulators responsive to AHL signals are LuxR family proteins. The LuxR family is composed of proteins about 250 amino acids in length and contains two functional domains. The first domain is the amino-terminal domain involved in AHL-binding and the second,a carboxy-terminal transcriptional regulation domain, which contains a helix-turn-helix DNA-binding motif. The LuxR-type proteins bind their cognate AHL with a stoichiometric ratio of one molecule of protein to one molecule of AHL. This complex dimerizes and specifically recognizes a dyad symmetric sequence which is the *lux* box. This sequence is located in the promoter regions of the target genes (Qin *et al.,* 2000; Urbanowski *et al.,* 2004; Zhu and Winans 2001). The acyl-HSLs signals have been termed "folding switches" due to the fact that upon binding the transcriptional activators fold into the stable conformation. Further stabilization may occur by dimerization of the signal-monomer complex. Most of the LuxR-type proteins are activators with the LuxR protein/DNA interaction fostering recruitment of the RNA polymerase to the promoter and therefore activate transcription of the downstream gene(s) (Schuster *et al.,* 2004; Kiratisin *et al.,* 2002). In contrast, several LuxR-type proteins act as transcriptional repressors whose DNA-binding activity is reduce by interaction with AHL signals (Minoque *et al.,* 2002; Horng *et al.,* 2002).

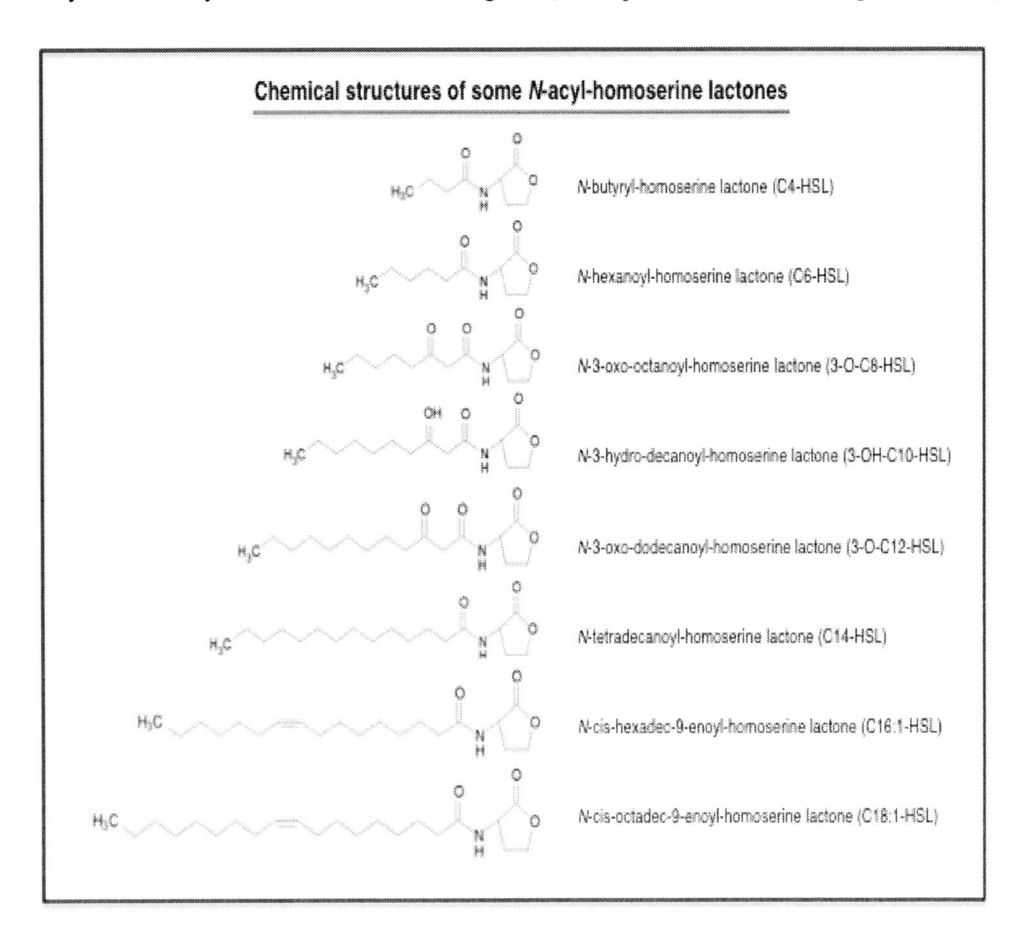

Figure 2. Chemical structures, names and abbreviations of some acyl-HSLs.

The crystal structure of the *Agrobacterium tumefaciens* TraR and *Pseudomonas aeruginosa* LasR activators has been solved.Most of the conserved amino acid residues in LuxR orthologs are involved in acyl-HSL signal recognition and DNA binding.TraR binds DNA at inverted repeat regions in promoter regions of regulated genes (Vannini *et al.,* 2002; Zhang *et al.,* 2002).The crystal structure of the LasR-ligand-binding domain to 3-oxo-C12-HSL has also been revealed (Bottomley *et al.,* 2007; Gould *et al.,* 2004).The LasR and TraR proteins have ligand-binding pockets adapted to their respective cognate acyl-HSLs, 3-oxo-C12-HL and 3-oxo-C8-HL, respectively, and the LasR binding pocket is larger (~670 Å^3)

than the TraR pocket (~440 Å^3).The larger acyl-HSL binding pocket of LasR thus may promote its ability to interact with larger long-chain non-cognate acyl-HSLs (Bottomley *et al.*, 2007).

3. BIOREPORTERSBASED ON REPORTER GENES *luxCDABE* FOR ACYL-HSL DETECTION

The first approach to determine if a bacterial strains contains a *luxI/luxR* cell-to-cell communication system often requires the detection of acyl-HSLs.Each bacterial strain may produce one or more acyl-HSL signals.This can depend upon the growth medium used and environmental conditions present at the time of bacterial growth.Usually one of the acyl-HSLs produced by a bacterial strain will be specific for the cognate LuxR receptor and thus bind and activate transcription of the target genes at the lowest concentrations in comparison to the other signals produced by the bacterium under investigation.The LuxR proteins binds and respond optimally to the acyl-HSL produced by the cognate LuxI-family protein, this fosters selectivity.Most of the acyl-HSL bioreporters detect a narrow range of acyl-HSLs signal molecules.Therefore, investigators typically employ multiple bioreporters each harboring a different acyl-HSL-dependent LuxR-type receptor.

Live whole-cell bioreporters containing sensing elements or receptors offer greater flexibility than pure chemical sensing elements.This is particularly relevant when taking into consideration of the compounds bioavailability, its affect on living systems and synergistic or antagonistic responses to a mixture of biota.In addition, bioreporters have the advantage of tailored engineering, which has been demonstrated by computational design, construction, and integration of a high-affinity ligand binding proteins into biosensors (Looger *et al.*, 2003).

A number of whole-cellacyl-HSL-dependent specific bioreporters have been developed over the last twenty years.This development was made possible due to the large number of acyl-HSL QS systems identified and characterized (Steindler and Venturi, 2007).The first acyl-HSL-dependent specific bioreporters utilized the *luxCDABE* operon as the reporter genes and were coupled to the sensing of cognate acyl-HSL signals by its LuxR ortholog (Swift *et al.*, 1997;Winson *et al.*, 1998).Table 1 lists the whole-cell bacterial bioreporters containing a LuxR ortholog as an acyl-HSL-dependent receptor to trigger bioluminescence as the reporter phenotype.These bioreporters do not produce acyl-HSLs but harbor a recombinant construct containing a native LuxR-family protein together with a cognate promoter made up of the *luxI* synthase / LuxR ortholog responsive elements, which are both involved in controlling expression of the *lux* operon reporter gene (Figure 3A).

3.1. The Importance of a Control Strain

Toxicity of the samples can result in misinterpretation of the presence of acyl-HSLs in samples.This false response can be controlled-for by including in the experiment a near-isogenic bioluminescence control bioreporter (Figure 3B).A control bioreporter strain allows signal correction in response to reduced cell viability.When sample toxicity reduces luminescence signal of the bioluminescent control strain by more than 50% compared with untreated cells, the sample is considered too toxic for reliable quantification of the acyl-HSL activity, and the sample should be diluted or partitioned into an organic solvent to reduce the toxic effects of the analyte or matrix.In addition, the parallel use of a constitutive bioluminescent control strain allows output signal correction by taking into consideration any nonspecific effects on cell viability.

The use of a constitutive bioluminescence control strain allows for output signal correction for nonspecific stimuli in samples that enhance or disturb the cell growth and, as a consequence, cause nonspecific alternations in the bioreporter gene expression.For this purpose, the sample species / strain used to construct the acyl-HSL-dependent bioreporter is transformed with a plasmid capable of constitutively expressing the bioluminescence operon.The signal increase or decrease in this strain is a sign of nonspecific sample effects.If this strain is near-isogenic, it is assumed that the sample affects the acyl-HSL reporter strain to the same degree; the signal from the acyl-HSL reporter strain can be corrected for nonspecific sample effects by dividing it with the signal from the control strain.The inclusion of a background or constitutive control strain is critical – especially when measuring environmental samples, which often contain nutrients or toxic compounds.In addition, both types of control strains allow a quick evaluation of cell viability, eliminating the need to count cells.

3.2. Detection of Short-chain acyl-HSLs

Biosensors containing a cognate LuxR-family receptor sensitive for acyl-HSLs with carbon chains of equal to or less than eight have been constructed that report bioluminescence (Table 1).These utilize the LuxR of *V. fischeri* and AhyR of *Aeromonas hydrophyla*.The LuxR-based biosensors are harbored in *E. coli*, thus do not produce acyl-HSLs, but contain recombinant plasmids pSB401 (Winson *et al.*, 1998) or pHV2001⁻(Pearson *et al.*, 1994), both containing the *luxR* of *V. fischeri* and cognate *luxI* promoter controlling *luxCDABE* expression.These biosensors are most sensitive to cognate acyl-HSL, 3-oxo-C6-HSL but can detect quite well C6-HSL, 3-oxo-C8-HSL and C8-HSL.However, weak agonist activity was observed with C4-HSL or C10-HSL or other long-chain acyl-HSLs (Winson *et al.*, 1998).An additional bioreporter strain,*E. coli* VJS533 (pHB2001⁻), based on the LuxR receptor and the *V. fischeri* ES114 *lux* regulon with an inactivated *luxI* in pBR322 detects 3-oxo-C6-HSL, C6-HSL, 3-oxo-C8-HSL and C8-HSL (Pearson *et al.*, 1994). The AhyR-based biosensor, also harbored in *E. coli*, is sensitive to acyl-HSL, C4-HSL.This biosensor, JM109 (pSB536) contains the *ahyR* and the *ahyI* promoter fused to the *luxCDABE* (Swift *et al.*, 1997).It is has not been reported whether LuxR- or AhyR-based biosensors could detect 3-hydroxy-HSLs or long-chain acyl-HSL with or without double bonds in the acyl side chain.

Similar to JM109 (pSB536) the Ahmer laboratory constructed a near-isogenic pair of biosensors for C4-HSL detection (Lindsay and Ahmer, 2005). These biosensors are in the *E. coli* SdiA- background with the *rhlR*gene in the case of JLD271 (pAL101) and without *rhlR* in the case of JLD271 (pAL102), but both containthe cognate *rhlI* promoter fused to*luxCDABE* (Table 1). SdiA is a solo LuxR-type protein that can activate the *rhlI* promoter and thus interfering with C4-HSL detection (Lindsay and Ahmer, 2005). This control plasmid pair or set enables the *luxR* dependence of light production to be determined. Each plasmid was introduced into *E. coli* stain JLD271 carrying an inactivation of the native SdiA gene (Lindsay and Ahmer, 2005). This set of bioreporters strains lacking the *sdiA* gene and carrying pAL101 and pAL102 has worked well in our laboratory to detect C4-HSL at concentrations as low as 0.5 μM.

Table 1. Characteristics of some acyl-HSL-dependent specific*luxCDABE*bioreporters. The table shows the grouping of bioreporters by the cognate signal detected

Plasmid Sensor	Host Cell	Acyl-HSL Receptor / Regulator	Responsive Promoter Element for *luxCDABE*	Cognate acyl-HSL	Reference
pSB536	*E. coli*JM109	AhyR	P*ahyI*	C4-HSL	Swift *et al.*, 1997
pSB406	*E. coli*JM109	RhlR	P*rhlI*	C4-HSL	Winson *et al.*, 1998
pAL101	*E. coli sdiA* mutant, JLD271	RhlR	P*rhlI*	C4-HSL	Lindsay and Ahmer, 2005
pHV2001⁻	*E. coli* VJS533	LuxR	P*luxI*	3-O-C6-HSL	Pearson *et al.*, 1994
pSB401	*E. coli* JM109	LuxR	P*luxI*	3-O-C6-HSL	Winson *et al.*, 1998
pSB403, wide-host range	*E. coli*, & *C. violaceum*	LuxR	P*luxI*	3-O-C6-HSL	Winson *et al.*, 1998
pCF218, pMV26	*A. tumefaciens*	TraR	P*traI*	3-O-C8-HSL	Chambers *et al.*, 2005; Sokol *et al.*, 2003
pSB1075	*E. coli*JM109	LasR	P*lasI*	3-O-C12-HSL	Winson *et al.*, 1998: Savka et al., 2010
pAL105	*E. coli sdiA* mutant, JLD271	LasR	P*lasI*	3-O-C12-HSL	Lindsay and Ahmer, 2005
pAL102	*E. coli sdiA* mutant, JLD271	No LuxR homolog, control for pAL101	P*rhlI*	none	Lindsay and Ahmer, 2005
pAL106	*E. coli sdiA* mutant, JLD271	No LuxR homolog, control for pAL105	P*lasI*	none	Lindsay and Ahmer, 2005

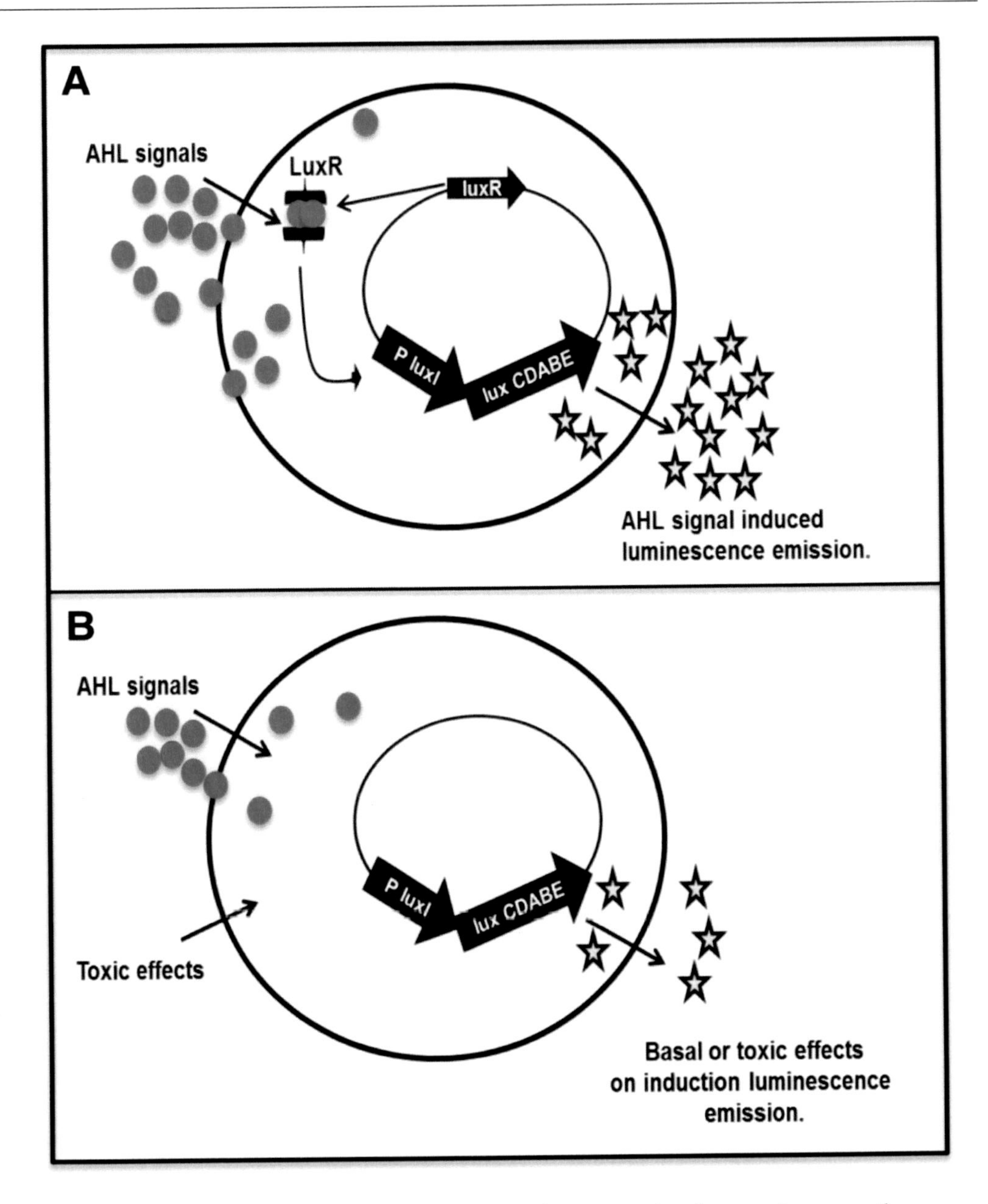

Figure 3. Generalized construction of specific bacterial bioreporters that detect and measure the presence of acyl-HSLs.A,Bioreporter with functional *luxR*-type receptor for acyl-HSL detection.B,Near-isogenic background control bioreporter partner identical to theconstruct shown in A but without a *luxR*-type receptor gene for acyl-HSL detection.This background-type control bioreporter serves to present basal-level bioluminescence output and can be used to determine if the samples being tested contain toxic metabolites that are inhibitory to the bioreporter metabolism or viability.Near-isogenic bioreporter pairs that only differ by the presence or absence of a LuxR acyl-HSL-dependent receptor are exemplified by the JLD271 (pAL – series) constructed in the Ahmer laboratory (Lindsay and Ahmer, 2005) (Table 1).

3.3. Detection of Long-chain Acyl-HSLs

Bioluminescence bioreporters for acyl-HSLs with carbon chains greater than eight have also be constructed with native cognate LuxR-family receptors.A biosensor for the detection of C10-HSL, 3-oxo-C10-HSL, C12-HSL and cognate acyl-HSL 3-oxo-C12-HSL uses the LasR receptor of *P. aeruginosa*.This biosensor based in*E.coli* JM109 contains pSB1075 and contains the *lasR* gene and cognate *lasI* gene promoter controlling *luxCDABE* expression (Winson *et al.,* 1998).

We recently reported on the relative sensitivity of a LasR-dependent *E. coli*-based bioluminescent biosensor JM109 (pSB1075) (Winson *et al.,* 1998) for rapid detection of twelve non-cognate long-chain acyl-HSLs with and without double bonds as well as those oxygenated at the third carbon as compared to the LasR-cognate acyl-HSL signal, 3-oxo-C12-HL (Pearson *et al.,* 1994, Savka *et al.,* 2010).In addition, we showed utility of biosensor JM109 (pSB1075) to detect long-chain acyl-HSLs encoded by *Agrobacterium vitis* from culture supernatants (Savka *et al.,*2010).Some of these findings are presented in Figure 4B and C

The Ahmer laboratory also constructed a pair of near-isogenic bioreporters that are different only in the presence or absence of the *lasR* gene.This set of strains allows for the determination of the light production which is dependent upon the presence of the LasR receptor.This pair of reporter plasmids, pAL105 and pAL106, were introduced into the *sdiA*$^-$ *E.coli* strain JLD271 (Lindsay and Ahmer, 2005).This set of bioreporter strains lacking the *sdiA* gene and carrying pAL105 and pAL106 has worked well in our laboratory to detect 3-oxo-C12-HSL at concentrations as low as 10 nM.

3.4. Bioluminescent Bioreportersfor Broad Range Detection of Acyl-HSLs

Consistent with the versatility and sensitivity of three*Agrobacterium*-based biosensors, one with a native *traR* (Shaw *et al.,* 1997; Farrand *et al.,* 2002), another with*traR* expressed from the *tetR* promoter (Zhu *et al.,* 1998) and the third with *traR* under the control of the phage T7 promoter (Zhu *et al.,* 2003), and all carrying a suitable responsive promoter (*traG, traI*) fuse to the *lacZ* gene, the Sokol laboratory constructed a *traI::luxCDABE* bioluminescent reporter fusion (Sokol *et al.,* 2003; Chambers *et al.,* 2005).The *traI::luxCDABE* reporter plasmid (pMV28) was moved into an *Agrobacterium* host carrying the *tetR::traR* fusion in plasmid pCF218 to overproduce the TraR acyl-HSL receptor protein.In this *Agrobacterium*-based bioluminescent bioreporter, TraR binds acyl-HSLs present in the sample preparation or culture extracts resulting in an acyl-HSL-TraR complex that subsequently dimerizes and binds the *lux* box sequence within the *traI* promoter on pMV28, triggering the transcriptional activation of the *lux* operon and thus light output (Bernier *et al.,* 2008).Bioreporter *A. tumefaciens* A136 (pCF218) (pMV26) detects all 3-oxo-AHLs with acyl side chains ranging from 4 to 12 carbons with the greatest sensitivity to cognate signal, 3-oxo-C8-HSL.Other acyl-HSL detected by A136 (pCF218) (pMV26) includes the unsubstituted acyl-HSL with acyl side chains ranging from 6 to 14 carbons (Chambers *et al.,* 2005; Bernier *et al.,* 2008).In addition, biosensor*A. tumefaciens* NTL4 (pZLR4) which contains the TraR receptor can detect 3-hydroxy acyl-HSLs including C6-OH-, C7-OH- C8-OH-, and C10-OH-HSL (Shaw *et al.,* 1997; Khan *et al.,* 2007); thus, *A.*

tumefaciens A136 (pCF218) (pMV26) should likewise andwith light production as the measurable output.

3.5. Use of Acyl-HSL-specific Bioluminescent Bioreporters

Bacterial light-producing bioreporters for acyl-HSL detection can be used in different ways.Initially it is simple to test an unknown strain that can be grown on solid media by streaking the unknown adjacent to the bioreporter (T-streak) or to streak the unknown strain across the biosensor (Cross-streak).In this scenario a light gathering instrument or photographic paper such as a charge-coupled device (photon / CCD camera or auto radiographic paper, respectively) to detect light production in *lux* operon-based bioreporters is required.Bioluminescence will be the strongest at the meeting point of the test strain and the bioreporter and a gradient will usually be evident as the two strains deviate from each other.The use of the bioreporter in this approach requires that both the unknown test strain and the bioreporter strain both grow well on the solid media being used.It is sometimes necessary to mix two different media to enable robust growth of both strains in Cross- or T-streak assays (Swift *et al.,* 1997; Gan *et al.,* 2009).

A second and widely used approach is coupling acyl-HSL signal separation with thin layer chromatography (TLC) on C_{18} reverse-phase plates followed by bioreporter detection.For this approach, acyl-HSL are extracted with acidified ethyl acetate after harvesting the late exponential phase culture supernatant or after resuspended from agar medium plates (Lowe *et al.,* 2009).This organic extraction is then dried and the residue is resuspended in smaller volumes of acidified ethyl acetate to concentrate the extract up to 100-fold.The TLC plates are then loaded with sample extracts and with different acyl-HSL standards.After chromatography, the plate is overlaid with a soft-agar suspension of the acyl-HSL bioreporter.Each acyl-HSL migrates with a characteristic mobility and results in a spot shape of response by using a CCD camera or another device to image light emission spatially.Alternatively, auto radiographic paper can be placed against the bioreporter overlay to detect and accumulate the photons being emitted by the bioreporter.The 3-oxo-acyl-HSLs produce a tear-drop spots whereas the alkanoyl-acyl-HSL and 3-OH-HSL migrate and produce circle-shaped spots (Shaw *et al.,* 1997).This coupled separation by TLC followed by detection by bioluminescence bioreporter give a direct visual measure of the numbers of acyl-HSL signals produced by an unknown bacterial isolate.The chromographic properties can be used to assign tentative structure by comparing their relative migration (Rf-values) for the unknown with Rf values for acyl-HSL standards.An example of this approach is shown in Figure 4A.This approach can be made semi-quantitative by loading different concentrations of the synthetic acyl-HSL signal(s) on the TLC plate (Farrand *et al.,* 2002; Fletcher *et al.,* 2007).The limitations of this approach include that no single bioreporter strain can detect the whole range of known acyl-HSLs and that each bioreporter exhibits a wide range of sensitivities for different acyl-HSL signals.To unequivocally identify the acyl-HSL structure the determination of spectroscopic properties are required using mass spectrometry (MS) and nuclear magnetic resonance spectroscopy (NMR) (Cataldi *et al.,* 2007; Gould *et al.,* 2006;Kirwan *et al.,* 2006; Savka *et al.,* 2010).

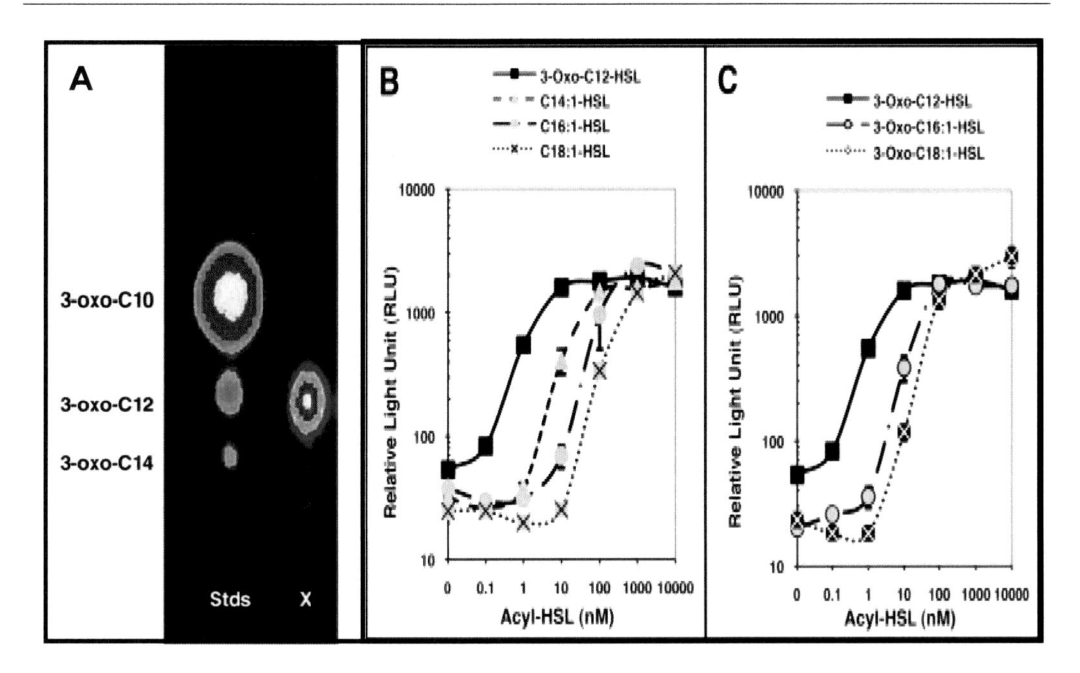

Figure 4.Analysis of acyl-HSL signals byseparation by thin layer chromatography coupled with detection by bioreporter JM109 (pSB1075), and by dose-response curves to long-chain acyl-HSL signals using bioluminescence bioreporter JM109 (pSB1075).A, TLC analysis of plant-produced long-chain acyl-HSLs.Solvent extraction of aqueous cell extracts from transgenic plant leaf tissue expressing the lasI gene from P. aeruginosa (Scott et al., 2006) were assayed with bioreporter JM109 (pSB1075) followed by detection with a CCD camera.Long-chain acyl-HSLs standards are present in lane Stds on the left and include 3-oxo-C10-HSL, 3-oxo-C12-HSL and 3-oxo-C14-HSL.The transgenic plant extract shown in lane X contains an acyl-HSL species with mobility characteristics of 3-oxo-C12-HSL.B and C, JM109 (pSB1075) dose-response, as measured by relative light units (RLU) or bioluminescence, to long-chain acyl-HSL signals with unsubstituted 3rd carbons in the acyl side chain but containing a double bond in the side chain (B) and long-chain acyl-HSL signals with 3-oxo-substitutions and double bonds in the acyl side chain (C), both sets of acyl-HSLs in comparison to the cognate acyl-HSL, 3-oxo-C12-HSL (Savka et al., 2010).

Finally, the bioluminescent bioreporters are particularly handy in quantification analysis of acyl-HSLs.For this purpose a luminometer or CCD camera is needed to quantify relative light units in each sample.This approach is useful to study differences in signal production among different strains (Li *et al.*, 2006;Lowe *et al.*, 2009) and to study the regulation of acyl-HSL synthesis (Gan *et al.*, 2009).In order to quantify accurately, one must determine the minimal concentration of acyl-HSL required for a response in a particular bioreporter as well as the concentration of acyl-HSL necessary for saturated response.This approach will enable to plot the linear dose response.From the linear response data a regression formula can be derived to semi-quantitatively estimate acyl-HSL concentrations in unknown samples.It is important to remember that unknown samples typically contain multiple acyl-HSL and the dose response curve is typically derived from one commercially prepared acyl-HSL.Typically, the generation of standard dose response curves using 5 to 200 nM concentrations of a particular acyl-HSL.This range of acyl-HSL can vary greatly and it is dependent on the sensitivity of the particular bioreporter and the acyl-HSL signal structure.An example of the use of bioluminescent bioreporter JM109 (pSB1075) for signal quantification is shown in Figure 4B and 4C.

3.6. Design of *in vitro* Cell-free Bioassay for Acyl-HSLDetection

The development of a bioluminescence acyl-HSL signal cell-free assay has been recently reported (Kawaguchi *et al.*, 2008).This successful effort decreases the assay time of whole-cell detection bioassays by reducing the cell conditioning growth phase and the time length of incubation, and eliminates the need to adjust the light output relative to bioreporter cell density at the end of the bioassays in each determination.This rapid and simplified assay overcomes the limitations of whole cell assays has been developed using cell lysates from a culture of the *A. tumefaciens* bioreporter carrying the *tetR::traR* fusion in plasmid pCF218 to overproduce the TraR acyl-HSL receptor protein and a β-galactosidase gene, *lacZ*, driven by the *traI* promoter in pCF372 (Zhu *et al.*, 1998).This bioassay is designed to report the product of β-galactosidase activity on the chromogenic substrate X-Gal that can be readily detected at 635 nm by spectrophotometry (an insoluble blue product).In this study, Kawaguchi and coworkers (2008) have also improved the sensitivity of this cell-free approach by the development of a luminescent approach using Beta-Glo (Promega).This luminescence cell-free bioassay increased the sensitivity by 10-fold incomparison to the absorbance-based bioassay.This system should be suitable for high-throughput screening of bacterial isolates, extracts and metagenomic libraries for the presence of acyl-HSL signals.

3.7. Design of Portable Bioreporter for Acyl-HSL Detection

A bioreporter for the detection of acyl-HSL signals that can serve as a simple and portable field kit has recently been developed in the laboratory of Sylvia Daunert (Struss *et al.*, 2010).This acyl-HSL bioreporter uses the LasR receptor protein whose cognate signal is 3-oxo-C12-HSL with the corresponding *lasI* promoter fragment to regulate the output of the *lacZ* gene that encodes β-galactosidase activity only in the presence of exogenous acyl-HSLs.This enzyme activity is directly proportional to the exogenous acyl-HSL in the environment.Theβ-galactosidase activity, a hydrolase enzyme, has been widely used in reporter systems for over 20 years and thus a variety of substrates are available for measuring its enzymatic activity.These include substrates to enable the output to be colorimetric, electrochemical, fluorescent or chemiluminescent (Struss *et al.*, 2010).This system has been demonstrated in a filter paper-strip bioreporter system to detect acyl-HSL from saliva of healthy and diseased individuals in the context of both colorimetric and chemiluminescent methods.This system should also be successful in monitoring acyl-HSL signals in environmental samples in remote field locations.

3.8. Potential Steps in the Detection Pathway to Increase Sensitivity and Future Directions

In whole-cell based bioreporters, the diffusion of a substrate and product through the cell wall or membrane results in a slower response than that of enzyme-based sensors.Thisdiffusion limitation could be improved by physical, chemical, and enzymatic approaches (also see review; van der Meer *et al.*, 2004).

Diffusion of Acyl-HSLs

The whole-cell QS acyl-HSL signal bioassays as describe in this review measures only the acyl-HSL signals that are bioavailable analytes.This refers to analytes that are able to enter the host cell.Efflux experiments using the two QS acyl-HSL systems in *P. aeruginosa* confirmed these observations.*P. aeruginosa*utilizes two LuxI / LuxR circuits to regulate many virulence factors.Pearson and coworkers (Pearson *et al.,* 1999) showed that the cognate signal for the RhlI/R system, C4-HSL is freely permeable; however, the long-chain acyl-HSL signal for the LasI/R system is not freely diffusible.These results where based on the*mexA-mexB-oprM*-encoded efflux pump shown to be involved in active efflux of long-chain 3-oxo-C12-HSL.From this work, it appears that the length and/or degree of substitution of theacyl side chain in acyl-HSL signals determine if they are freely diffusible or are subject to transmembrane pumps.Further work in this area is needed to confirm if additional long-chain acyl-HSLs use transmembrane systems for transport.

Use of LuxR Solos asReceptorsin acyl-HSL Biosensors

In most cases, *luxI* and *luxR* homologs are found at same genetic locus (Case *et al.,* 2008).However, in the last five years a number of LuxR solo have been described.The use of solo LuxR-type receptors and corresponding responsive promoter regions in acyl-HSL bioreporter constructions has not been investigated for wide spread applications.Thus, the use of certain *luxR* solos in combination with responsive promoter elements could be useful to construct bioreporters that detect a single or many acyl-HSLs.A review on LuxR-family solos sensor/regulators of acyl-HSLs lists twelve functionally characterized LuxR solos (Subramoni and Venturi, 2009).Of the twelve listed, two bind a single acyl-HSL, one binds two acyl-HSLs, three bind 5 or more acyl-HSLs, one is ligand independent, while the remaining five LuxR solos bind a yet to be discovered molecule.Further work with LuxR solos proteins will uncover candidates for used in specific recombinant bioreporter constructs for detection of certain acyl-HSLs.

Bioengineered LuxR Receptors

The use of genetically modified acyl-HSL receptors proteins with altered acyl-HSL binding sites could increased sensitivity of certain signals.Furthermore, the design of modified promoter / operator regions on reporter proteins, partners to the acyl-HSL-dependent receptors may enable binding to the RNA polymerase differentially and thereby altering the biosensor response.Two mutants in the TraR receptor, T129A and T129V, detected unsubstituted acyl-HSLs with the same affinity as the cognate signal of TraR, 3-oxo-C8-acyl-HSL, which contains a carbonyl substitution at the 3rd carbon of the acyl side chain (Chai and Winans, 2004).Likewise, in the studies to alter the LuxR receptor, Collins and coworkers, used a dual positive-negative selection system to first direct a mutation leading to "promiscuous" functions and second to "respecialization" function (Collins *et al.,* 2005; 2006).In these works, LuxR mutants, produced by error-prone PCR followed by DNA shuffling, led to LuxR variants that showed 100-fold increase sensitivity to C8-acyl-HSL while remaining a wild –type or greater response to cognate signal, 3-oxo-C8-acyl-HSL.In addition, these mutants also responded to long-chain acyl-HSL, C14-acyl-HSL (Collins *et al.,* 2005).In the second study, one of the LuxR mutants from the previous study, named LuxR-G2E, was used in error-prone PCR, and one variant with a single additional amino acid change was identified and named LuxR-G2E-R67M.In gene activation studies with LuxR cognate signal, 3-oxo-C6-HSL,LuxR-G2E-R67M was near background levels and this contrasted significantly with both the LuxR-G2E and wild-type LuxR variants.In addition,gene activation with C6-HSL, showed significant induction with LuxR-G2E-R67M and this contrasted with wild-type LuxR, with a 100-fold difference required for minimal activation.Similar, but reciprocal, differencesbetween LuxR-G2E-R67M and LuxR were found in gene activation experiments using the long-chain acyl-HSL, C10-acyl-HSL (Collins *et al.,* 2006).

These LuxR-type variants can be used in bioreporter constructions for both broad range detection of acyl-HSLs and for new specific and sensitive detection of certain acyl-HSL signal(s).Such variants will also supplement the tool box of transcription factors that can be important in biotechnology and synthetic biology applications of engineering cell-cell communications in genetic circuit designs, tissue engineering, biofabrication and targeted gene therapies (Choudhard and Schmidt-Dannert, 2010; Young and Alper, 2010).

Other than the luxCDABEgenes as the Bioluminescence Reporter

Bioluminescence proteins such as luciferase and aequorin, rather than fluorescent proteins may be preferable, particularly in whole cell biosensors. Bioluminescence does not suffer from the typical limitations of fluorescence, such as high background, photobleaching and need for a light source. On the other hand, fluorescent proteins, such as the green fluorescent protein and its variants, can be detected in real time without addition of any substrates or cofactors.

Nevertheless, fluorescent proteins lack the sensitivity of luciferases and other bioluminescent proteins, mainly due to high fluorescence background.Bioluminescent proteins can be detected at attomole levels, allowing ultrasensitive detection of the target analytes and monitoring of the physiological phenomena under investigation.It also enables the analyses of small-volume samples, which should facilitate the development of miniaturized and high-throughput assays (Daunert*et al.,* 2000;van der Meer and Belkin, 2010).

CONCLUSION

Besides the use of bioluminescent bioreporters for acyl-HSL detection, many additionalbacterial biosensors have been constructed that use different reporter outputs.These include:β-galactosidase, ice nucleation protein, green fluorescent protein, the production of purple pigment, violacein, and ablue-green pigment produced in response to C4-HSL by *P. aeruginosa*strain CGMCC 1.860 (McClean *et al.,* 1997;Llamas *et al.,* 2004; DeAngelis *et al.,* 2007;Ling *et al.,* 2009;Steindler and Venturi, 2007; Steide *et al.,* 2001; Yong and Zhong, 2009).However, bacterial bioreporter systems based on bioluminescent *luxCDABE* reporter genes under the transcriptional control of the LuxR-acyl-HSL complex allow powerful, rapid, sensitive, cost-effective and quantitative detection and quantification of acyl-HSL cell-to-cell communication signals.

REFERENCES

Asad S, Opal SM. 2010. Bench-to-bedside review: Quorum sensing and the role of cell-to-cell communication during invasive bacterial infection. *Critical Care* 12(6):1-11.

Atkinson S, Williams P. 2009. Quorum sensing and social networking in the microbial world.*J R Soc Interface* 6:959-978.

Bassler BL, Wright M, Silverman MR. 1994. Multiple signaling systems controlling expression of luminescence in *Vibrio harveyi*: sequence and function of genes encoding a second sensory pathway.*Mol Microbiol* 13(2):273-286.

Bernier SP, Beeston AL, Sokol PA. 2008. Detection of *N*-acyl homoserine lactones using a *traI-luxCDABE*-based biosensor as a high-throughput screening tool.*BMC Biotechnology,* 8:59doi:10.1186/1472-6750-8-59

Bottomley MJ, Muraglia E, Bazzo R, Carfi A. 2007. Molecular insight into quorum sensing in the human pathogen *Pseudomonas aeruginosa* from the structure of the virulence regulator LasR bond to its autoinducer. *J Biol Chem282*:13592-13600.

Case RJ, Labbate M, Kjelleberg S. 2008. AHL-driven quorum-sensing circuits: their frequency and function among the Proteobacteria. *ISME J* 2:345-349.

Cataldi RRI, Gianco G, Palazzo L, Quaranta V. 2007.Occurrence of *N*-acyl-L-homoserine lactones in extracts of some Gram-negative bacteria evaluated by gas chromatograph-mass spectrometry. *Analytical Biochemistry*361:226-235.

Chai Y, Winans SC. 2004. Site-directed mutagenesis of a LuxR-type quorum-sensing transcriptional factor: alternation of autoinducer specificity. *Mol Microbiol* 51:765-776.

Chambers CE, Visser MB, Schwab U, Sokol PA. 2005. Identification of *N*-acylhomoserine lactones in mucopurulent respiratory secretions from cystic fibrosis patients. *FEMS Microbiol Lett* 244:297-304.

Choudhary S, Schmidt-Dannert C. 2010. Applications of quorum sensing in biotechnology.*Appl Microbiol Biotechnol* 86:1267-1279.

Collins CH, Leadbetter JR, Arnold FH. 2006. Dual selection enhances the signaling specificity of a variant of the quorum-sensing transcriptional activator LuxR. *Nature Biotechnology* 24(6):708-712.

Collins CH, Arnold FH, Leadbetter JR. 2005.Direct evolution of *Vibrio fischeri* LuxR for increased sensitivity to a broad spectrum of acyl-homoserine lactones.*Mol Microbiol* 55:712-723.

Dalzell DJ, Slte S, Aspichueta E, de la Sota A, Etxebarria J Gutierrez M, Hoffmann CC, Sales D, Obst U, Christofi N. 2002. A comparison of five rapid direct toxicity assessment methods to determine toxicity of pollutants to activated sludge. *Chemosphere* 47:535-545.

Daunert S, Barrett G, Feliciano JS, Shetty RS, Shrestha S, Smith-Spencer W. 2000. Genetically engineering whole-cell sensing systems: coupling biological recognition with reporter genes. *Chem. Rev.* 100:2705-2738.

DeAngelis KM, Firestone MK, Lindow SE. 2007.Sensitive whole-cell biosensor suitable for detecting a variety of *N*-acyl homoserine lactones in intact rhizosphere microbial communities. *Appl Environ Microbiol* 73:3724-3727.

Duerkop BA, Ulrich RL, Greenberg RP. 2007. Octanoyl-homoserine lactone is the cognate signal for *Burkholderia mallei* BmaR1-BmaI1 quorum sensing. *J Bacteriol* 189:5034-5040.

Farrand SK, Qin Y, Oger P. 2002. Quorum-sensing system of *Agrobacterium* plasmids: Analysis and Utility. *Methods Enzymol*358:452-484.

Ferluga S, Bigirimana J, Hofte M, Venturi V. 2007. A LuxR homologue of *Xanthomonas oryzae pv. oryzae* is required for optimal rich virulence. *Mol Plant Pathol* 8:529-538.

Fletcher MP, Diggle SP, Camara M, Williams P. 2007. Biosensor-based assays for PQS, HHQ and related 2-alkyl-4-quinolone quorum sensing signal molecules. *Nature Protocols 2(*5):1254-1262.

Fuqua C. 2006. The QscR quorum-sensing regulon of *Pseudomonas aeruginosa:* an orphan claims its identify. *J Bacteriol.* 188:3169-3171. (doi:10.1128/JB.188.9.3169-3171.2006)

Fuqua C, Winans S. 1999.Signal generation in autoinduction systems: synthesis of acylated homoserine lactones by LuxI-type proteins, pages 211-230, In: *Cell-Cell Signaling in Bacteria*, G. M. Dunny and S. C. Winans (Eds.), ASM Press.

Gan HM, Buckley L, Szegedi E, Hudson AO,Savka MA. 2009.Identification of an rsh gene from a *Novosphingobium* sp. necessary for quorum-sensing signal accumulation. *J Bacteriol*191:2551-2560.

Gould TA, Herman J, Krank J, Murphy RC, Cook DM, Churchill MEA. 2006.Specificity of acyl-homoserine lactone synthases examined by mass spectrometry. *J Bacteriol* 188:773-783.

Gould TA, Schweizer HP, Churchill MEA. 2004.Structure of the *Pseudomonas aeruginosa* acyl-homoserine lactone synthase LasI.*Mol Microbiol* 53:1135-1146.

Hanzelka BL, Parsek MR, Val DL, Dunlap PV, Cronan JE Jr, Greenberg EP. 1999. Acylhomoserine lactone synthase activity of the *Vibrio fischeri* AinS protein.*J Bacteriol 181*(18):5766-5770.

Horng YT, Deng SC, Daykin M, Soo PC, Wei JR, Luh KT, Ho SW, Swift S. Lai HC, Williams P. 2002. The LuxR family protein SpnR functions as a negative regulator of N-acylhomoserine lactone-dependent quorum sensing in *Serratia marcescens.Mol. Microbiol* 45(6):1655-1671.

Jiang Y, Camara, M, Chhabra SR, Hardie KR, Bycroft BW, Lazdunski A, Salmond GPC, Stewart GSAB and Williams P. 1998. In vitro biosythesis of the *Pseudomonas aeruginosa* quorum-sensing signal molecule *N*-butanoyl-L-homoserine lactone.*Mol. Microbiol.*28:193-203.

Kawaguchi T, Chen YP, Norman RS, and Decho AW. 2008. Rapid screening of quorum-sensing signal *N*-acyl homoserine lactones by an in vitro cell-free assay. *Appl Environ Microbiol* 74:3667-3671.

Khan SR, Herman J, Krank J, Serkova NJ, Churchill ME, Suga H, Farrand SK. 2007. *N*-(3-hydroxyhexanoyl)-l-homoserine lactone is the biologically relevant quormone that regulates the *phz* operon of *Pseudomonas chlororaphis* strain 30-84.*Appl Environ Microbiol* 73(22):7443-7455.

King JMH, PM DiGrazia, B Applegate, R Burlage, J. Sanseverino, P. Dunbar, F. Larimer, and GS Sayler. 1990. Rapid, sensitive bioluminescent reporter technology for naphthalene exposure and biodegradation. *Science* 249:778-781.

Kiratisin P, Tucker KD, Passador L. 2002. LasR, a transcriptional activator of *Pseudomonas aeruginosa* virulence genes, functions as a multimer.*J Bacteriol* 184(17):4912-4919.

Kirwan JP, Gould TA, Schweizer HP, Bearden SW, Murphy RC,, and Churchill MEA. 2006. Quorum-sensing signal synthesis by the *Yersinia pestis* acyl-homoserine lactone synthase YspI. *J. Bacteriol.*188:784-788.

Laue BE, Jiang Y, Chhabra SR, Jacob S, Stewart GS, Hardman A, Downie JA, O'Gara F, Williams P. 2000. The biocontrol strain *Pseudomonas fluorescens* F113 produces the *Rhizobium* small bacteriocin, N-(3-hydroxy-7-cis-tetradecenoyl)homoserine lactone, via HdtS, a putative novel *N*-acylhomoserine lactone synthase.*Microbiology* 146(10):2469-2480.

Lee HJ, Gu MB. 2003. Construction of *a sodA::luxCDABE* fusion *Escherichia coli*:comparison with a *katG* fusion strain through their responses to oxidative stresses. *Appl Microbiol Biotechnol* 60:577-580.

Lee JH, Lequette Y, Greenberg EP. 2006. Activity of purified QscR, a *Pseudomonas aeruginosa* orphan quorum-sensing transcriptional factor. *Mol Microbiol* 59:602-609. (doi:10.1111/j.1365-2958.2005.04960.x)

Lequette Y, Lee JH, Ledgham F, Lazdunski A, Greenberg EP. 2006. A distinct QscR regulon in the *Pseudomonas aeruginosa* quorum-sensing circuit. *J Bacteriol* 188:3365-3370. (doi:10.1128/JB.188.9.3365-3370.2006)

Li Y, Gronquist MR, Hao G, Holden MR, Eberhard A, Scott RA, Savka MA, Szegedi E, Sule S, & other authors. 2006.Chromosome and plasmid-encoded *N*-acyl-homoserine lactones produced by *Agrobacterium vitis*wild type and mutants that differ in their interactions with grape and tobacco. *Physiol. Mol. Plant Path.*67:284-290.

Ling EA, Ellison ML, Pesci EC.2009. A novel plasmid for detection of *N*-acyl homoserine lactones. *Plasmid* 62(1):16-21.

Lindsay A, Ahmer BMM. 2005. Effect of *sdiA* on biosensors of *N*-acylhomoserine lactones. *J Bacteriol 187*:5054-5058.

Llamas I, Keshavan N, Gonzalez JE. 2004. Use of *Sinorhizobium meliloti* as an indicator for specific detection of long-chain *N*-acyl-homoserine lactones.*Appl. Environ. Microbiol.* 70:3715-3723.

Loh J, Pierson EA, Pierson LS, Stacy S, Chatterjee A. 2002.Quorum sensing in plant-associated bacteria.*Curr Opin Plant Biol* 5:285-290.

Looger LL, Dwyer MA, Smith JJ, Hellinga HW. 2003. Computational design of receptor and sensor proteins with novel functions. *Nature*423:185-190.

Lowe N, Gan HM, Chakravartty V, Scott R, Szegedi E, Burr TJ, and Savka MA.2009. Quorum-sensing signal production by *Agrobacterium vitis* strains and their tumor-inducing and tartrate-catabolic plasmids. *FEMS Microbiol Lett* 296:102-109.

McClean KH, Winson MK, Fish L, et al. 1997. Quorum sensing and *Chromobacterium violaceum*: exploitation of violacein production and inhibition for detection of *N*-acylhomoserine lactones. *Microbiology* 143:3703-3711.

Meighen EA. 1991. Molecular biology of bacterial bioluminescence. Microbiol Rev 55:123-142.

Milton DL, Chalker VJ, Kirke D, Hardman A, Camara M, Williams P. 2001. The LuxM homologue VanM from *Vibrio anguillarum* directs the synthesis of N-(3-hydroxyhexanoyl)homoserine lactone and *N*-hexanoylhomoserine lactone.*J. Bacteriol* 183(18):3537-3547.

Minogue TD, Wehland-von Trebra M, Bernhard F, von Bodman SB. 2002. The autoregulation role of EsaR, a quorum-sensing regulator in *Pantoea stewartii* ssp. *stewartii:* evidence for a repressor function. *Mol. Microbiol.* 44:1625-1635.

Mitchell RJ and Gu MB. 2004. An *Escherichia coli* biosensor capable of detecting both genotoxic and oxidative damage. *Appl Microbiol Biotechnol* 64:46-52.

More MI, Finger LD, Stryker JL Fuqua C, Eberhard A and Winans SC. 1996. Enzymaticsynthesis of a quorum-sensing autoinducer through use of defined substrates. *Science* 272:1655-1658.

Norman A, Hansen LH, Sorensen SJ. 2005. Construction of a ColD cda promoter-based SOS-green fluorescent protein whole-cell biosensor with higher sensitivity toward genotoxic compounds than constructs based on recA, umuDC or sulA promoters. *Appl Environ Microbiol* 71:2338-2346.

Ortori CA, Atkinson S, Chhabra SR, Camara M, Williams P, and Barrett DA.2006.Comprehensive profiling of *N*-acylhomoserine lactones produced by *Yersiniapseudotuberculosis* using liquid chromatography coupled to hybrid quadrupole-linear ion trap mass spectrometry. *Anal Bioanal Chem387*:497-511.

Parsek MR and Greenberg EP. 1999. Quorum-sensing signals in development of *Pseudomonas aeruginosa* biofilms. *Biofilms* 310:43-55.

Pearson JP, Van Delden C, Iglewski BH. 1999. Active efflux and diffusion are involved in transport of *Pseudomonas aeruginosa* cell-to-cell signals. *J Bacteriol* 181:1203-1210.

Pearson JP, Gray KM, Passador L, Tucker KD, Eberhard A, Iglewski BH, Greenberg EP. 1994. Structure of the autoinducer required for expression of *Pseudomonas aeruginosa* virulence genes. *Proc. Natl. Acad. Sci.* USA 91:197-201.

Preston S, Coad N, Townsend J, Killham K Paton GI. 2002. Biosensing the acute toxicity of metal interactions: are they additive, synergistic or antagonistic? *Environ Toxicol Chem* 2000. 19:775-780.

Qin Y, Luo ZQ, Smyth AJ, Gao P, Beck von Bodman S, Farrand SK. 2000. Quorum-sensing signal binding results in dimerization of TraR and its release from membranes into the cytoplasm.*EMBO J* 19(19):5251-5221.

Ramey BE, Koutsoudis M, von Bodman SB, Fuqua C. 2004. Biofilm formation in plant-microbe associations.*Curr Opin Microbiol* 6:602-9.

Sagi E, Hever N, Rosen R, Bartolome AM, Premkumar JF, Ulber R, Lev O, Scheper T, Belkin S. 2003. Fluorescence and bioluminescence reporter functions in genetically modified bacterial biosensors. *Sens Actuators* B 90:2-8.

Savka MA, Le TP and Burr JT. 2010. LasR receptor for detection of long-chain quorum-sensing signals: Identification of *N*-acyl-homoserine lactones encoded by the *avsI* locus of *Agrobacterium vitis*. *Current Microbiology;* EPub. Ahead of Print, June 1, 2010.

Schuster M, Urbanowski ML, Greenberg EP. 2004. Promoter specificity in *Pseudomonas aeruginosa* quorum sensing revealed by DNA binding of purified LasR.*Proc Natl Acad Sci USA* 101(45):15833-15839.

Scott RA, Weil J, Le PT, Williams P, Fray RG, von Bodman SB, Savka MA. 2006. Long- and short-chain plant-produced bacterial *N*-acyl-homoserine lactones become components of phyllosphere, rhizosphere and soil. *Mol. Plant-Microbe Interact.*19:227-239.

Shaw PD, Ping G, Daly SL, Cha C, Cronan JE, Rinehart KL, Farrand SK. 1997. Detecting and characterizing *N*-acyl-homoserine lactone signal molecules by thin-layer chromatography.*Proc. Natl. Acad. Sci. USA* 94:6036-6041.

Smith S, Wang J-H, Swatton JE, Davenport P, Price B, Mikkelsen H, Stickland H, Nishikawa K, Gardiol, N. & other authors.2006.Variations of a theme: diverse *N*-acyl-homoserine lactone-mediated quorum sensing mechanisms in Gram-negative bacteria. *Science* Progress 89:167-211.

Sokol PA, Sajjan U, Visser MB, Gingues S, Forstner J, Kool C. 2003. The CepIR quorum-sensing system contributes to the virulence of *Burkholderia cenocepacia* respiratory infections. *Microbiology* 149:3649-3658.

Steidle A, Sigl K, Schuhegger R, Ihring M, Schmid M, Gantner S, Stoffels M, Riedel K, Givskov M, and other authors.2001. Visualization of *N*-acylhomoserine lactone-mediated cell-cell communication between bacteria colonizing the tomato rhizosphere.*Appl. Environ. Microbiol.* 67:5761-5770.

Steindler L, Venturi V. 2007. Detection of quorum-sensing *N*-acyl homoserine lactone signal molecules by bacterial biosensors.*FEMS Microbiol. Lett.* 266:1-9.

Stevens AM,Greenberg EP. 1997. Quorum sensing in *Vibrio fischeri:* Essential elements for activation of the luminescence genes. *J. Bacteriol.* 179:557-562.

Struss M, Pasini P, Ensor CM, Faut N, Daunert S. 2010. Paper strip whole cell biosensors: a portable test for the semiquantitative detection of bacterial quorum sensing molecules. *Anal Chem.* 82:4457-4463.

Subramoni S, Venturi V. 2009. LuxR-family 'solos': bachelor sensors/regulators of signaling molecules. *Microbiology* 155:1377-1385.

Swift S, Karlyshev AV, Fish L, Durant EL, Winson MK, Chhabra SR, Williams P, Macintyre S, Stewart GS. 1997. Quorum sensing in *Aeromonas hydrophila* and *Aeromonas salmonicida:* identification of the LuxRI homologs AhyRI and AsaRI and their cognate *N*-acylhomoserine lactone signal molecules. *J Bacteriol* 179:5271-5281.

Taga ME, and Bassler BL. 2003.Chemical communication among bacteria.*Proc. Natl. Acad. Sci. USA* 100:14549-14554.

Ulitzur S, Lahav T, Ulitzur N. 2002. A novel and sensitive test for rapid determination of water toxicity.*Environ Toxicol* 17:291-296.

Unge A, Tombolini R, Molbak L, Jansson JK. 1999. Simultaneous monitoring of cell number and metabolic activity of specific bacterial populations with a dual *gfp-luxAB* marker system. *Appl Environ Microbiol* 65:813-821.

Urbanowski ML, Lostroh CP, Greenberg EP. 2004. Reversible acyl-homoserine lactone binding to purified *Vibrio fischeri* LuxR protein.*J. Bacteriol.* 186(3):631-637.

Van der Meer JR and Belkin B. 2010. Where microbiology meets microengineering: design and applications of reporter bacteria. *Nature Reviews Microbiology* 8:511-522.

Van der Meer JR, Tropel D, Jaspers MCM. 2004.Illuminating the detection chain of bacterial bioreporters. *Environ. Microbiol.* 6:1005-1020.

Vannini A, Volpari C, Gargioli C, Murageli E, Cortese R, De Francesco R, Neddermann P, Marco SD.2002. The crystal structure of the quorum sensing protein TraR bound to its autoinducer and target DNA. *EMBO J.* 21:4393-4401.

Williams, P. 2007.Quorum sensing, communication and cross-kingdom signaling in the bacterial world.*Microbiology* 153:3923-3938.

Wilson T and Hasting W. 1998.Bioluminescence. *Annu Rev Cell Dev Biol* 14::197-230.

Winson, MK, Swift S, Fish L, Throup JP, Jorgensen F, Chhabra SR, Bycroft BW Williams P, Stewart GS. 1998. Construction and analysis of *luxCDABE*-based plasmid sensors for investigating *N*-acyl-homoserine lactone-mediated quorum sensing.*FEMS Microbiol. Lett.* 163:185-192.

Yong Y-C, Zhong J-J. 2009. A genetically engineered whole-cell pigment-based bacterial biosensing system for quantification of *N*-butyryl homoserine lactone quorum sensing signal. *Biosensors and Bioelectronics* 25:41-47.

Young E, Alper H. 2010. Synthetic biology: tools to design, build, and optimize cellular processes. *J. Biomed and Biotechnol Article ID 130781,* 12 pages. doi: 10.1155/2010/130781.

Zhang R, Pappas T, Brace JL, Miller PC, Qulmassov T, Molyneaus JM, Anderson JC, Bashkin JK, Winans SC, other authors.2002.Structure of bacterial quorum-sensing transcriptional factor complexed with pheromone and DNA. *Nature*417:971-974.

Zhu J, BeaberJW, More MI, Fuqua C, Eberhard A, Winans SC. 1998. Analogs of the autoinducer 3-oxooctanoyl-homoserine lactone strongly inhibits activity of the TraR protein of *Agrobacterium tumefaciens*. *J Bacteriol.*180:5398-5405.

Zhu J, Chai Z, Zhong Z, Li S, Winans SC.2003.*Agrobacterium* bioassay strain for ultrasensitive detection of *N*-acyl-homoserine lactone-type quorum-sensing molecules: detection of autoinducers in *Mesorhizobium huakuii*.*Appl. Environ. Microbiol.*69:6949-6953.

Zhu J, Winans SC. 2001. The quorum-sensing transcriptional regulator TraR requires its cognate signaling ligand for protein folding, protease resistance, and dimerization.*Proc Natl Acad Sci USA* 98(4):1507-1512.

In: Bioluminescence
Editor: David J. Rodgerson, pp. 71-95

ISBN 978-1-61209-747-3

Chapter 4

THE USE OF BACTERIAL BIOSENSORS TO DETERMINE THE TOXICTY OF NANOMATERIALS AND OTHER XENOBIOTICS

R. I. Dams**, *A. Biswas, A. Olesiejuk, M. Gómez, M. M. Murcia, N. Christofi and T. Fernandes
Pollution Research Unit, Edinburgh Napier University,
Merchiston Campus, Edinburgh EH10 5DT, Scotland UK

ABSTRACT

It is widely accepted that the current and planned increase in commercial development of nanotechnology will lead to pronounced releases of nanomaterials (NM) in the environment. In this study we aim to evaluate the toxicity of silver nanoparticles and some chlorinated phenol compounds by using a common sludge bacterium, Pseudomonas putida (originally isolated from activated sludge) as biosensor in bioluminescence assays. The results of the toxicity testing were expressed as IC50 values which represents the amount of toxicants required to reduce light output by 50% and calculated using a statistical program that was developed in-house. The bacterium carrying a stable chromosomal copy of the lux operon (luxCDABE) was able to detect toxicity of ionic and particulate silver over short term incubations ranging from 30-240 min. The IC50 values obtained at different time intervals showed that highest toxicity (lowest IC50) was obtained after 90 min incubation for all toxicants and this is considered the optimum incubation for testing. The data show that ionic silver is most toxic followed by nanosilver particles with microsilver particles being least toxic. Release of nanomaterials is likely to have an effect on the activated sludge process as indicated by this study using a common sludge bacterium involved in biodegradation of organic wastes. Among the chlorinated phenol compounds, 3,5 dichlorophenol showed the highest toxicity and phenol being the least toxic. Among the hydroxylated compounds

* Author for correspondence. E-mail address: rosemedero@yahoo.co.uk Phone/fax: +44 0131-455-2291

tested, benzoquinone showed to be the most toxic compound. Understanding the fate and transport of phenolic compounds and transformation products in removal treatment plants is relevant in evaluating the impact of discharge in the environment and in developing a program for pollution control. Our results demonstrate that the use of a bacterial biosensor as P. putida BS566::luxCDABE, native from a highly polluted environment, provides a robust, early warning system of acute toxicity which could lead to process failure. This strain is suitable for toxicity monitoring in a highly polluted industrial waste water treatment streams and it has the potential to inform on upstream process changes prior to their impacting upon the biological remediation section.

Keywords: bacterial biosensors, Pseudomonas putida, silver nanoparticles, wastewater treatment, phenol compounds, ecotoxicity

1. Introduction

Bacterial Biosensors

Bioluminescence is associated with the emission of light by living microorganisms and it plays a very important role in real time process monitoring. Biosensors are able to monitor very low levels of chemicals, can work in complex matrices and have very fast response times. Furthermore, it requires short times of analysis and provide low cost of assays, portable equipment, real-time measurements. When the reporter lux gene is fused to a promoter regulated by the concentration of a compound of interest, this concentration can be quantitatively analyzed by detecting the bioluminescence intensity. On the other hand, when the reporter is fused to promoters that are continuously expressed as long as the organism is alive and metabolically active, it will be proper for evaluating the total toxicity of contaminant.

Recombinant bioluminescent biosensing cells have been widely used to determine the toxicity of several xenobiotics as herbicide paraquat (Belkin et al., 1997), phenolic compounds (Ben-Israel et al., 1998; Gu and Choi, 2001; Boyd et al., 2001) and its degradation byproducts (Choi and Gu, 2003) and recently to test the toxicity of oxide nanoparticles to bacteria and crustaceans (Heinlaan et a., 2008) to protozoa (Mortimer et al., 2003), to microalgae (Aruoja et al., 2009), to protozoa and crustaceans in natural waters (Blinova et al., 2010).

Nanoparticles Toxicity

The knowledge of the ecotoxicology of nanoparticles (NP) to bacteria and other microbes is still limited, even though some manufactured nanoparticles (NP; materials with three dimensions between 1-100nm; Bridges et al., 2007) such as silver and titanium, which are constantly released into the environment, are known as antibacterial agents. As such it is essential that technology that fully assesses their effects on natural microbial entities, and on biogeochemical cycling in the environment, is available. Clearly there is a concern that these novel materials could be released into the environment, and that there may be releases from products that are in current use. However, we are only just starting to explore their ecotoxicology and environmental chemistry.

The increased commercial development of nanotechnology has the potential release of such nanoparticles in the environment, however, knowledge on their fate and effects on biota in natural habitats is still lacking, particularly using real systems such aquatic and marine waters and terrestrial systems. A particular knowledge gap relates to the effect of NP on the ubiquitous microbial community of the planet that is responsible for the recycling of organic material and the continued fertility of aquatic and terrestrial systems. Bacteria perform many critical roles required for normal ecosystem function and productivity, such as carbon and nitrogen cycling. Toxicity of nanoparticles to bacteria has caused many concerns. Interactions between bacteria and NPs may provide us with more information about the impact of NPs once released into the ecosystem. At the same time, bacteria as single cell organisms are good test models to study the NPs' toxicity and to know how the NPs affect the cell/organism function.

Although is not clear which mechanism is responsible for the toxic effect of nanoparticles on bacterial cells, quite a lot of information is available for nano-sized particles due to their use as bactericides. Sondi and Salopek-Sondi (2004) observed aggregates of silver nanoparticles in dead cells of E. coli using scanning electron microscopy. Other studies have observed the citotoxicity of iron-based nanoparticles toward Escherichia coli where the authors (Auffan et al., (2008) found out that nanoparticles containing zerovalent iron were toxic to the bacteria probably due to the generation of reactive oxygen species.

Sinha et al., (in press) studied the effect of silver and zinc nanoparticles towards mesophilic and halophilic bacteria cells. They observed that nanotoxicity was more pronounced on halophilic Gram negative bacteria than mesophilic bacteria cells. Also, it was observed by the authors (Sinha et al., 2010) that Gram positive halophilic bacteria showed a resistance towards both silver and zinc nanoparticles.

Among the oxide nanoparticles tested (aluminium, silicon, titanium and zinc) Jiang et al., (2009) found out that zinc was the most toxic to Bacillus subtilis, E. coli and Pseudomonas fluorescens. Clearly, toxicity is also species dependent and it can be associated with characteristics of certain bacterial species. For instance, Kim et al., (2007) observed a severe inhibitory effect on the growth of yeast and E.coli and a mild effect on Staphylococcus aureus. Siejak and Frackowiak (2007) have showed that the intensity of fluorescence of Cianobacterium synechocytis strongly diminished in the presence of 30nm silver nanoparticles.

One of the consequences of the increased utilization of nanomaterials is the release of these materials to the environment and consequently the amounts of nanowaste should increase. Source of nanomaterials maybe from the collection systems in municipalities where huge amounts of nanoparticles are release daily. The amount of nanoparticles reaching soil and natural waters depends on the fraction of wastewater that is effectively treated. A simplified pathway has been described by Bystrzejewska-Piotrowska et al., (2009) from nanotechnology to nanowaste, where unbound nanoparticles may be released from disposed or residue paints, cosmetics or pharmaceuticals, other nanoproducts, textiles or coatings may form composites. Toxic ions (from metallic nanoparticlers such as silver) may be released due to leaching.

Silver nanoparticles is well known for its antibacterial activity and have become one of the nanomaterials most used in consumer product (Maynard and Michelson, 2006), it is considered as the most prevalent of engineered materials (Rejeski and Lekas, 2008) and is likely that nanowaste should increase and therefore enter the wastewater treatment plants. Blaser et al. (2008) pointed out that silver released to wastewater is incorporated into sewage sludge and may spread further on agricultural fields where will mainly stay in the top layer of soils (Hou et al. 2005). Land filling of sewage sludge may allow silver to leach into subsoil and groundwater. Estimation of silver load in sewage sludge and its microorganisms growth inhibition has been predicted. Blaser et al., (2008) predicted that an expected silver concentration in sewage treatment plant range from 2 μg L-1 to 18 μg L-1. Shafer et al., (1998) reported a range of ~ 2 to 4 μg L-1 of silver in sewage treatment plants treating common wastewater and a much higher load from industrial discharges (from 24 to 105 μg L-1). Choi et al., (2008) have demonstrated that Ag-NP have an inhibitory effect on the growth of autotrophic and heterotrophic bacteria in wastewater treatment plants. In these studies, Choi et al., (2008) found out that silver ion was the most toxic species (among silver nanoparticles, silver ions and silver loride colloids) to inhibit heterotrophic growth. Therefore, its release into the environment as waste has arisen serious concern. Certainly, nanomaterials do not behave in the same way as normal waste ans therefore there is a need for a more suitable and fast system which allows a early warning about the entrance of such nanomaterials into the waste water treatment. The current standard wastewater treatment technologies might not be suitable for removing NMs from the effluents, thus providing an escape route for dissolved chemicals as it has been postulated from Leppard et al., 2003. Recently, Westerhoff et al., (2008) demonstrated the wastewater treatment systems' inability to remove NMs from drinking water, therefore implying the potential presence of NMs in drinking water with an exposure pathway to humans.

The bioluminescence bioassay here performed is able to offer this possibility where in manner of minutes or few hours is possible to monitor the presence of these nanoparticles, specially silver nanoparticles, the object of this study.

Phenol Compounds

Chlorophenols are ranked as severe environmental pollutants and are important intermediates in the chemical industry (Fallmann et al., 1999) and have been used as biocides or as starting components for pesticide production (McAllister et al.,1996; Litchfield and Rao,1998; Chi and Huang, 2002). Highly chlorinated phenols have been listed as priority pollutant by the US Environmental Protection Agency (Keith and Telliard,1979). Chloro-substituted phenols and cresols are known to be toxic to humans as well as to aquatic life (Keith and Telliard, 1979). They often occur in industrial wastewaters and solid waste leachates causing their recalcitrance to biological treatments. Chlorinated compounds can be removed from the environment by physicochemical processes, biological degradation including various aerobic, anaerobic, and combined aerobic–anaerobic biological processes. And during their transformation, many byproducts are produced by partial degradation, which may result in increasing toxicity levels (Paton et al., 1997). Trichlorophenol and dichlorophenol have been found as intermediate compounds during dechlorination of PCP in horizontal-flow anaerobic immobilized biomass reactors (Damianovic et al. 2009) and PCP anaerobic degradation pathway (Madsen and Aamand ,1990). The production of intermediates from a partial dechlorination of PCP can generate even more toxic compounds as 3,4,5-trichlorophenol (TCP), which was found to be more toxic than PCP (Wu et al., 1989). The chemical 2,4,6-TCP has been considered one of the most toxic chlorophenols (Eker and Kargi, 2007).

Hydroxylated compounds such as catechol, hydroquinone, resorcinol and benzoquinone may be produced in the oxidative transformation of phenols (Fukushima et al., 2002; Gorska et al. 2009) and 4-chlorocathecol has been identified as intermediate in a *meta* cleavage of 4-chlorophenol (Hollender et al., 1997). Also, under ozonation treatment of PCP, benzoquinone and hydroquinone were identified as intermediate compounds (Hong and Zeng, 2002). Thus, removal of the original compounds can be considered as only a partial measure of the efficiency of the treatment.

It is generally accepted that in the assessment of toxicity of environmental pollutants, a range of bioassays are required to examine the variability in a toxic response involving organisms of different trophic levels. The same specifically applies to microorganisms and, bacteria in particular, where different species can respond differently to toxic chemicals. Whole microbial cell biosensors are now widely used as research tools in the testing of substances likely to elicit cytotoxic and genotoxic events, and, in the determination of bioavailability of chemicals (Belkin, 2007). There are many advantages using biosensors as monitoring very low levels of chemicals, working with complex matrices and have very fast response times (Dennison and Turner,1995). They embrace genetically engineered bacteria that have a toxicant detecting gene that is coupled with a reporter gene (e.g. luminescence gene such as lux or luc) capable of producing a detectable response on activation.

Table 1. Stock solutions' concentrations of chlorinated phenol and intermediate compounds

Compound	Stock solution (concentration g L^{-1})
Phenol	5 g/l
4-chlorophenol	1g/l
3,5 dichlorophenol	1g/l
2,4,6 trichlorophenol	1g/l
PCP	1g/l
4-chlorocatechol	1g/l
3,5 dichlorophenol	1g/l
Cathecol	0.500 g/l
Chlorocathecol	0.500 g/l
Benzoquinone	0.500 g/l
Chlorohidroquinone	0.500 g/l
Hydroquinone	0.500 g/l
Resorcinol	0.500 g/l
Chlororesorcinol	0.500 g/l
1,2,4 THB	0.500 g/l

In this ecotoxicity study we have focused on highly chlorinated phenols as pentachlorophenol, 2,4,6 trichlorophenol, 3,5 dichlorophenol and monosubstituted 4-chlorophenol and phenol due to their environmental relevance as pollutants. Phenolic compounds exert a toxic effect on microorganisms by disrupting energy transduction either by uncoupling oxidative phosphorylation or by inhibiting electron transport (Escher et al., 1996). Substituted phenols act by destroying the electrochemical proton gradient by transporting protons back across the membrane and/or inhibiting electron by binding to specific components of the electron transport chain (Escher et al., 1996).

As the luminescence of lux-marked bacterial biosensors depends on the products of the electron transport chain for light production, inhibition of this process due to chlorophenol toxicity will result in a decrease in light output.

In this study we performed ecotoxicological testing using a genetically-modified Pseudomonas isolated from a polluted, phenolic-rich, wastewater treatment system by transposon mutagenesis (Wiles et al., 2003). The Pseudomonas carries a stable chromosomal copy of the lux operon (luxCDABE) derived from Photorhabdus luminescens with continuous output of light.

Several compounds have been tested: chlorinated phenol compounds and some common by-products of degradation of phenolic compounds as hydroquinone, cathecol, chlorohydroquinone and benzoquinone. Silver materials in the ionic form, nanoparticles (35nm) and microparticles (0.6-1.6 microns).

2. Material and Methods

2.1. Chemicals

All of the chemicals used were analytical grade.

Phenol Compounds

Sodium Salt Pentachlorophenol (PCP) and Phenol were purchased from Sigma Chemical Company (USA); 3,5 dichlorophenol, 2,4,6 Trichlorophenol, and 4 Chlorophenol were purchased from Aldrich, Madison, Wis. 4-chlorocatechol (97%), resorcinol (99%), 4-chlororesorcinol (98%), chlorohydroquinone (85%) and p-benzoquinone (98%) were obtained from Aldrich. 1,2,4 trihydroxibenzene (THB), hydroquinone (99%), catechol (99%) were purchased from Sigma.

Stock solutions of PCP, 3,5 dichlrophenol, 4-chlorophenol, phenol, catechol, 4-chlorocatechol , resorcinol, 4-chlororesorcinol, hydroquinone, chlorohydroquinone, p-benzoquinone, 1,2,4 THB were freshly prepared in 0.85% saline solution. Stock solutions of 2,4,6 trichlorophenol was prepared in 0.2N NaOH solution. Working solutions were double diluted externally and then added to 96 well black microtitre plates used for the bioluminescence bioassays. Table I shows the stock solutions' concentrations used in the ecotoxicity tests.

Silver Species

Silver powder nanoparticles (average size: 35nm, 99.5% metal basis, spherical morphology and cubic crystallographic structure) and microparticles (average size: 0.6-1.6 microns) were obtained from Nanoamor (Nanostructured and Amorphous Materials Inc, Texas, USA).

Stock solutions of silver ion ($AgNO_3$), silver nanoparticles (Ag-NP) and silver microparticles (Ag-MP) were freshly made in 20 ml Universal flasks and placed into an ultrasonic bath (XB6 Grant Instruments Cambridge Ltd, UK) at 25 kHz, 250C for 30 minutes. Two fold dilutions were externally prepared of these solutions and added to the 96 well black microtitre plates (Sterling, Caerphilly, UK). Silver ion working solutions concentrations ranged from 0.23 μg L-1 to 2500 μg L-1. Silver nanoparticles solutions concentrations ranged from 0.195 mg L-1 to 200 mg L-1 and silver microparticles solutions concentrations ranged from 1.46 mg L-1 to 1500 mg L-1. Two different stabilisers were added to the nanoparticles and microparticles working solutions: 0.1% citric acid, 0.1% BSA (bovine serum albumin) and for comparison solutions with no stabiliser also were used. Previous studies have used BSA for stabilisation of ZnO nanoparticles (Brayner et al., 2006) and also as a carbon black nanomaterial stabiliser in suspension (Foucaud et al., 2007). Thus improving stability of particle suspension by decreasing the extent of particle agglomeration and settling over time (Foucaud et al., 2007). Citric acid is known to act as a chelating agent and may thus be able to also act to quench toxicity of dissolved metal toxicants and it has been found to reduce agglomeration of particles (Chu et al., 2002) thus improving particle distribution and response. All working solutions were light protected and used within 30 minutes of preparation. In order to verify no shading effects of the particles on the bacterial cells, light output was measured before and after addition of particles. No shading effects were determined.

2.2. Media and Growth Conditions

Organism

The bacterial strain used in this study, P. putida BS566::luxCDABE was constructed based upon chromosomal expression of the luxCDABE operon derived from an entomopathogenic nematode symbiont, Photorhabdus luminescens (Winson et al., 1998). Originally isolated from the treatment system (Whiteley et al., 2001), this reporter organism encompassing a dynamic xenobiotic sensing range, suitable for placement around an industrial processing system to monitor remediation in multiple compartments.. Cultures were grown in Luria-Bertani (LB) broth, containing 100 mgL-1 Kanamycin, overnight at 26oC with shaking (200rpm) to late exponential stage. Prior to the assay, cultures were diluted to approximately 107 cells/ml and regrown under the same conditions for two to three generations without Kanmycin. When OD (optical density) reached 0.2 (approximately 10 8 cell/ml) toxicity tests were carried out.

2.3. Biosensor Assay

Luminescence measurements were undertaken using a 96-well plate luminometer (FLUOstar Optima 1 and 2, BMG Labtech, UK) in 96 well black microtitre plates (Sterling, Caerphilly, UK) whereby each well contained bacterial inoculum and toxicant at the required concentration in 100 μL volumes, using an integration time of 1 s at a temperature of 28°C. Reading was taken at every 30 for 240 min. Control wells containing LB broth with P. putida

BS566::luxCDABE were run and changes in toxicity for the test systems are expressed as percentages of the control. Luminescence values were expressed in the instrument's arbitrary relative light units (RLU). The maximal response ratios were the highest ratios of luminescence in the sample-containing wells to luminescence in wells containing untreated cells determined during a specified period: 30, 90, 180 and 240 min (Belkin et., 1996).

Inhibitory concentration which represents 50% inhibition of light output (IC50) related to control was assessed for all toxicants tested. All experiments were run at least 3 times (most of 4 times) at different dates with different batches.

2.4. Calculation of IC_{50} (IC- Inhibitory Concentration) Values

The IC values are calculated using a statistical program that was developed in-house. The program fits a three parameter logistic model to the logarithm of the concentration by weighted least squares. The parameters are the initial response, the slope and the intercept. It is assumed that the response would decline to zero at sufficiently high concentrations. The initial response effectively uses the information from both the controls (if present) and low concentrations. The weights used are taken to be proportional to the fitted response but with adjustments for high and low responses; this is to protect against bias due to "hormesis" effects (stimulatory effects causing increased light output when challenged with low toxicant concentrations) and the effective omission of data respectively.

2.5. Data Analysis

For error analysis, all of the experiments were conducted three times on different plates. Data from at least eight wells were used for one concentration and coefficients of variation (CV) between independent assays were calculated using Microsoft Excell 97. Differences among treatments were tested using a two-way analysis of variance (ANOVA) to determine which treatments were statistically different ($P < 0.05$).

3. Results and Discussion

Phenol Compounds

Luminescence responses of the engineered strain P. putida BS566::luxCDABE was assessed using different concentrations of chlorinated phenol compounds of environmentally relevance and some intermediate compounds from their transformation. Luminescence activity was calculated as residual activity when compared to uninhibited samples, and expressed as a percentage of this control. This reporter was found to demonstrate a proportional decay in luminescence after incubation of 4h. Inhibitory concentration (IC_{50}) for each compound tested resulting in luminescence decay to 50% of the control luminescence was assessed.

Using the biosensor P. putida BS566::luxCDABE, IC_{50} values were determined (Table 2) for PCP, 2,4,6 trichlorophenol, 3,5 dicholorophenol, 4 chlorophenol, phenol, and the intermediate compounds: cathecol, chlorocathecol, benzoquinone, chlorohidroquinone, hydroquinone, chlororesorcinol, resorcinol and 1,2,4 THB.

Calculated coefficients of variation (CV) between independent assays were found to be between 2% and 16% (data not shown), which were lower than reported for another toxicity assays using Pseudomonas spp.(7 – 18%) (Brown et al. 1996; Paton et al. 1995), demonstrating the reproducibility and reliability of the developed toxicity assay.

Figure 1 shows the light output reduction by P. putida BS566::luxCDABE when challenged with phenol compounds: 4- Chlorophenol; 3,5 Dichlorophenol; 2,4,6 Trichlorophenol and PCP. Overall, the predicted toxicity of all phenolic compounds indicated that 3,5 dichlorophenol was statistically the most toxic compound (<0.5). The order of toxicity to P. putida was: 3,5 dichlorophenol > 2,4,6 trichlorophenol> PCP> 4 chlorophenol. Figure 2 shows the luminescence reduction by P. putida when challenged with phenol. This organism showed a high tolerance to phenol. The IC50 value for phenol after 30 min of exposure to this toxicant was 437 mg L-1 (Table 2), which is about 87 fold greater than those determined for 3,5 dichlorophenol. After 4 h of exposure the IC50 found for growth inhibition of P. putida was 225 mg L-1 , which was close to the value of 244 mg L-1 in a 6h test found by Slabbert (1986) in a 6h test using P. putida.

The IC_{50} values (Table 2) found for 2,4,6 trichlorophenol toxicity was 6.3 fold greater than that determined for phenol ($P<0.05$) after 30 min of exposure. After 90 min it dropped to ~3,5 and so remained until the end of the test (4h).

Table 2. IC_{50} (mg l^{-1}) values for light emission reduction by P. putida BS566::LUXCDABE after 30, 90, 180 and 240 min

	IC_{50} values (mg l^{-1})			
Time (min)	30	90	180	240
Phenol	437	237	218	225
4-chlorophenol	141	133	124	123
3,5 dicholorophenol	5	2,5	2,5	2,5
2,4,6 trichlorophenol	69	64	62	62
PCP	106	90	78	74
Chlorocathecol	160	115	93	66
Hydroquinone	159	393	325	300
Resorcinol	941	350	266	245
Cathecol	397	245	275	270
Benzoquinone	0.9	1.10	1.60	1.30
chlorohidroquinone	102	171	189	200
Chlororesorcinol	227	142	157	176
1,2,4 THB	41	79	88	122

Values presented are an average of at least 3 independent experiments carried out with different batches. Standard deviation between 2 and 16 %.

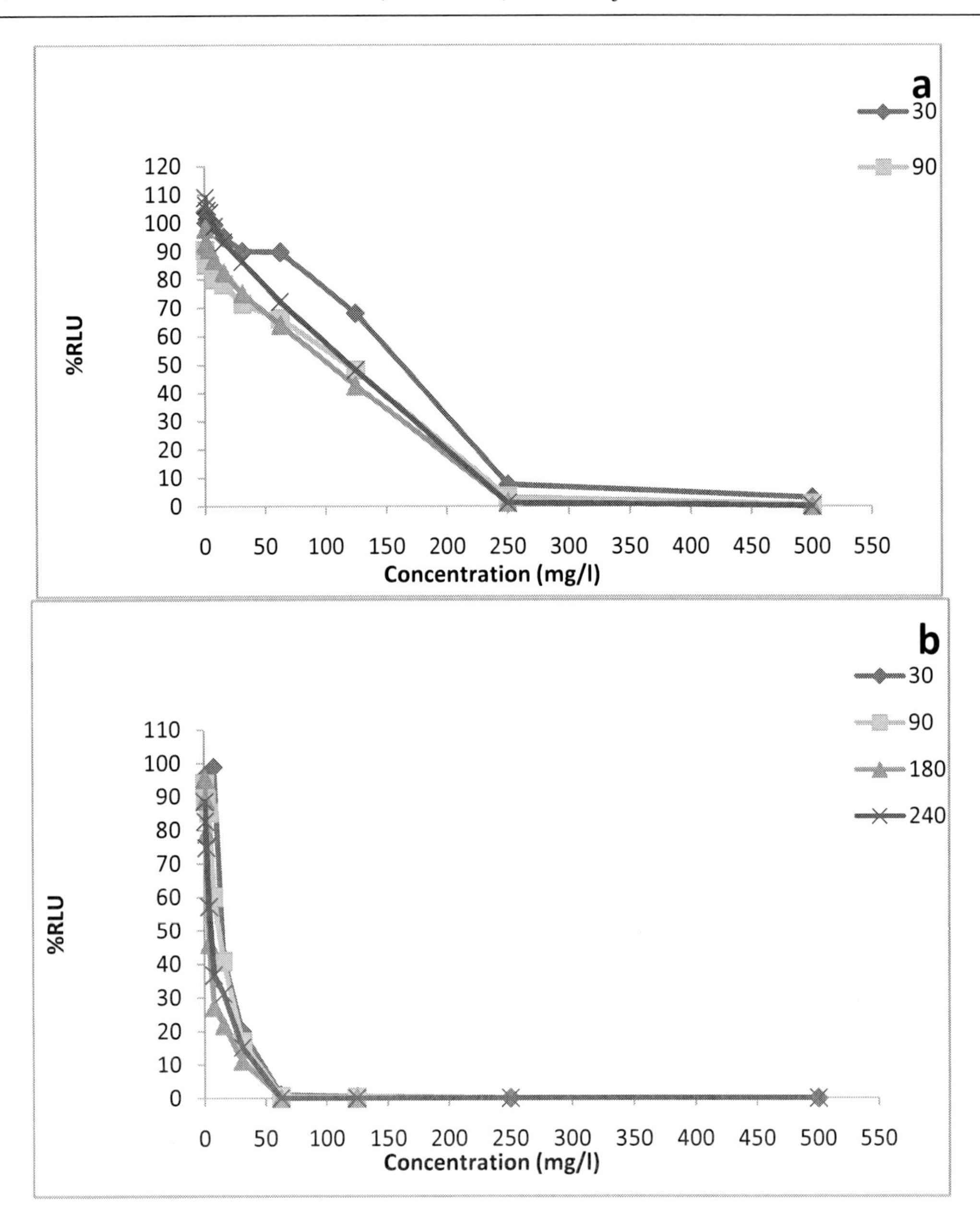

Figure 1 (continued).

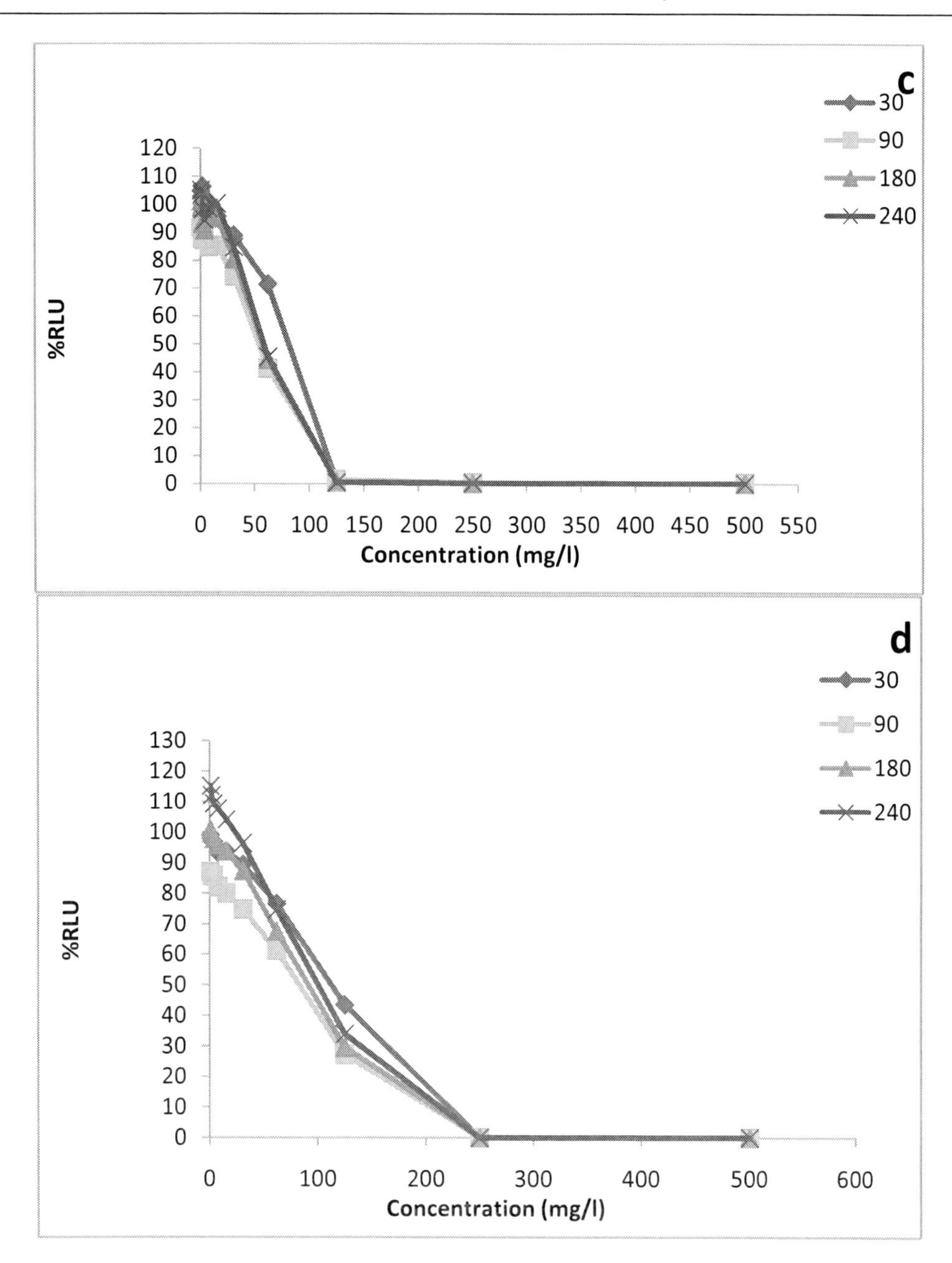

Figure 1. Light output reduction by P. putida BS566::luxCDABE when challenged with 4-Chlorophenol (a), 3,5 Dichlorophenol (b), mg L-1 2,4,6 Trichlorophenol (c) and PCP (d) after 30, 90, 80 and 240 min. Mean of three replicates with different batches. Coefficient of variation 12% among independent assays.

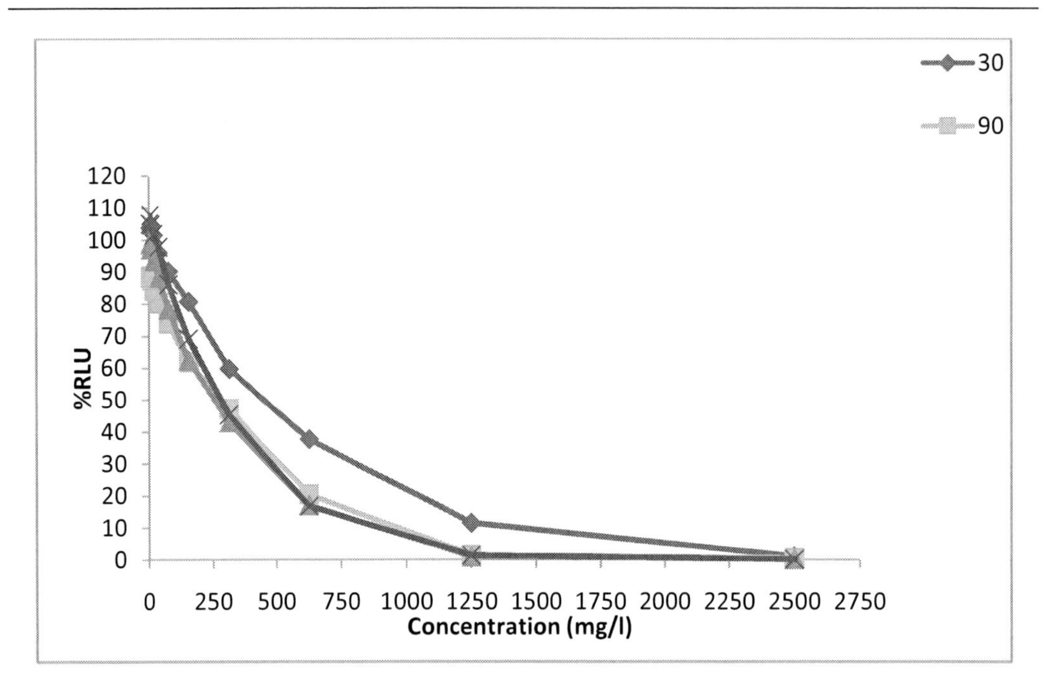

Figure 2. Light output reduction by P. putida BS566: luxCDABE when challenged with phenol after 30, 90, 80 and 240 min. Mean of three replicates with different batches. Coefficient of variation 7% among independent assays.

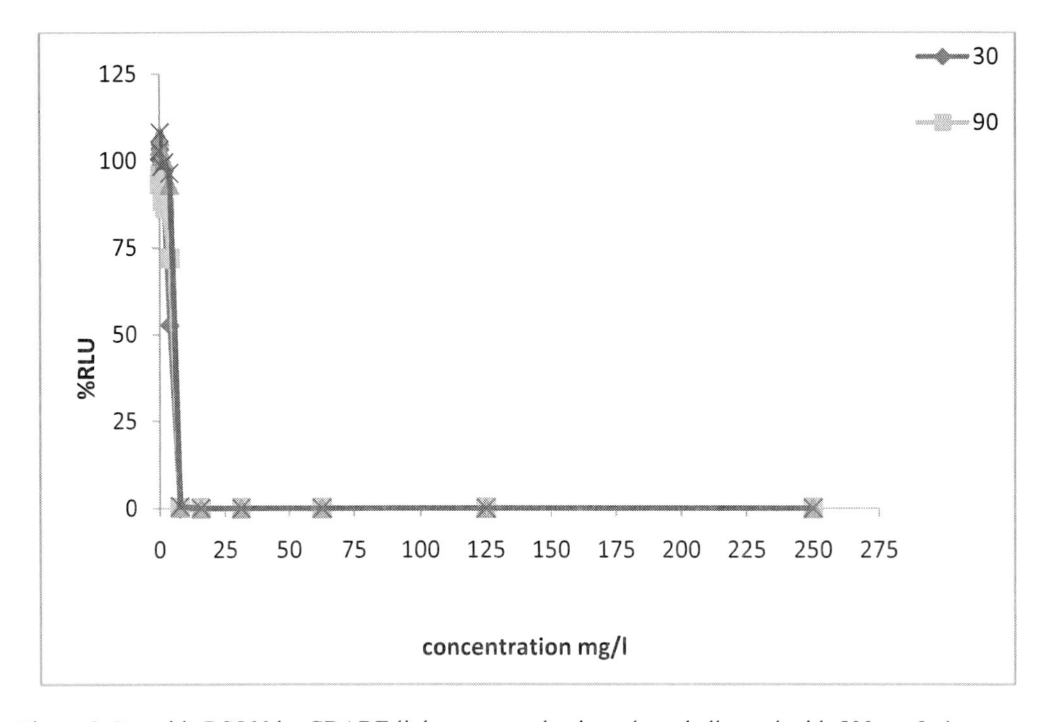

Figure 3. P. putida BS566:luxCDABE light output reduction when challenged with 500 mg L-1 Benzoquinone after 30, 90, 80 and 240 min. Mean of three replicates with different batches. Coefficient of variation 8 % among independent assays.

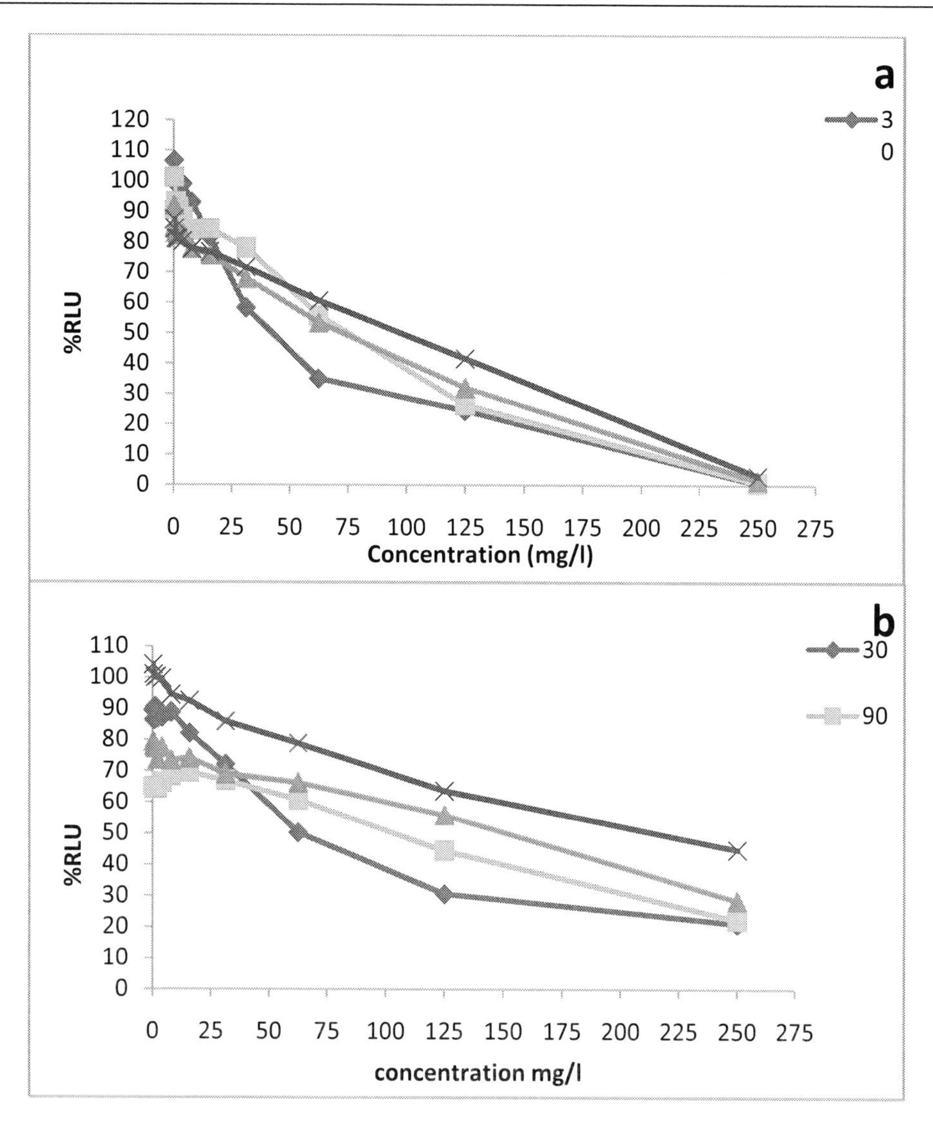

Figure 4. Light output reduction by P. putida BS566::luxCDABE when challenged with THB (a) and chlorohidroquinone after 30, 90, 80 and 240 min. Mean of three replicates with fifferent batches. Coefficient of variation 11% among independent assays.PCP and 4-chlorophenol toxicity was about 4 and 3 folder greater than that determined for phenol ($P<0.05$) after 30 min of exposure (Table 2) after 30 min of exposure, respectively.

Chlorinated phenol compounds as tri and di chlorophenol are common intermediates among removal treatments of phenolic compounds such as pentachlorophenol. For instance, the pentachlorophenol biodegradation pathway by Burkholderia cepacia AC1100 (Zeng and Wackett, 2008) and Sphingomonas chorophenolicum ATCC 39723 (Copley, 2000) showed that through oxidative cleavage of the original compound, hidroquinones, trichlorophenol and dichlorophenol are generated as intermediates. Our results showed that within a 30 min bioluminescent assay is possible to determine the high toxicity of the intermediate compounds when comparing with the original compound. Also, showed that monochlorophenol was less toxic than di, trichlorophenol and pentrachlorophenol. Other species of *Pseudomonas fluorescens* and *Burkholderia Rasc* c2 (Boyd et al., 2001) used as biosensors also showed a greater sensibility to dichlorophenols and trichlorophenols than those determined for monochlorophenols. Chlorine substitution adds electrophilicity, resulting in reactive toxicity. Additionally, increasing chlorination increases the dissociation constant, making phenol derivatives more acidic (Paulus, 1993) and more polar due the presence of a strong hydrogen bonding group (-OH) (Cronin and Schultz, 1998).

In this study, byproducts originated from 4-chlorophenol photodegradation studies formerly carried out in our laboratories (Gomez et al., 2010) such as phenol, cathecol,benzoquinone, resorcinol, hydroquinone and others were tested for its toxicity. Fenton's reaction generates hydroxyl-radicals (OH) by means of the reaction of H_2O_2 with ferrous ion, which produces a range of intermediates upon treatment of aromatic and chloroaromatic compounds, including chlorophenols. The elementary reaction produces cyclohexadienyl radicals, which upon stabilization yield the corresponding quinones (Bunce et al., 1997; Chen and Pignatello, 1997; Kang et al., 2002; Pera-Titus et al., 2004). Figure 3 shows the light reduction by *P.* putida BS566::luxCDABE when challenged with benzoquinone. In this study, among the byproductos tested, benzoquinone is far the most toxic compound. The order of toxicity is: benzoquinone> 1,2,4 THB> chlorohidroquinone> hydroquinone> chlorocathecol> chlororesorcinol> cathecol> resorcinol (Table 2).

Figures 4-6 show the light reduction by P. putida BS566::luxCDABE when challenged with THB, chlorohydroquinone, hydroquinone, chlorocathecol, chlororesorcinol, cathecol and resorcinol. Most of the byproducts showed no residual toxicity after 4 h of exposure (THB, chlorohidroquinone, hydroquinone, chlororesorcinol, cathecol and resorcinol. However, benzoquinone and chlorocathecol after 4 h of exposure showed a residual toxicity in the order of 94 and 2 fold greater (Table 2), respectively, when comparing with the original compound 4-chlorophenol. Hydroxylated and carbonyl compounds are typically involved in the contaminant degradation, and low molecular weight organic acids are usually detected as a residual of the oxidation (Mills and Hoffmann, 1993; Stefan and Bolton, 1998). Kim et al., (2007) suggested that the observed PCP post-treatment toxicity might be due to toxic PCP byproducts, which may include polychlorinated dibenzodioxins/furans,tetrachloro-*p*-benzoquinone, and other intermediates. Tetrachlorohydroquinone, tetrachlorobenzoquinone and tetrachlorocatechol have been reported previously as PCP intermediates of TiO2 photo-oxidation, UV photodegradation and ozonation (Benitez et al., 2003; Hong and Zeng, 2002; Hirvonen et al., 2000; Luo et al., 2008; Weavers et al., 2000). It has been reported that complete mineralization into carbon dioxide and inorganic is obtained only with prolonged treatment, even though the half-life of the original contaminant may be very short (Sundstrom et al., 1989; Mills and Hoffmann, 1993). In that context, removal of the original compounds is only a partial measure of the efficiency of the treatment. The degradation products formed in the oxidation processes are mainly low molecular weight oxygenated and less toxic, more biodegradable than the original compounds (Miller *et al.* 1988) as observed for most of the intermediate compounds in this study (Table II). Previous studies in our laboratories (Gomez et., 2010) showed that when 4-chlorophenol was treated with excimer lamp only (without hydrogen peroxide), toxic concentrations of several byproducts, including hydroquinone, benzoquinone, resorcinol, chlorohydroquinone, remained in the medium after treatment, while hydrogen peroxide improved the removal of photoproducts. The residual oxidation products can persist after treatment because of their low reactivity towards the oxidants (Arnold et al., 1995). Consequently, the potential formation of persistent or toxic intermediates should be evaluated if the partial oxidation fulfilled the criteria for discharge to the subsequent stage of the treatment process or into the environment.

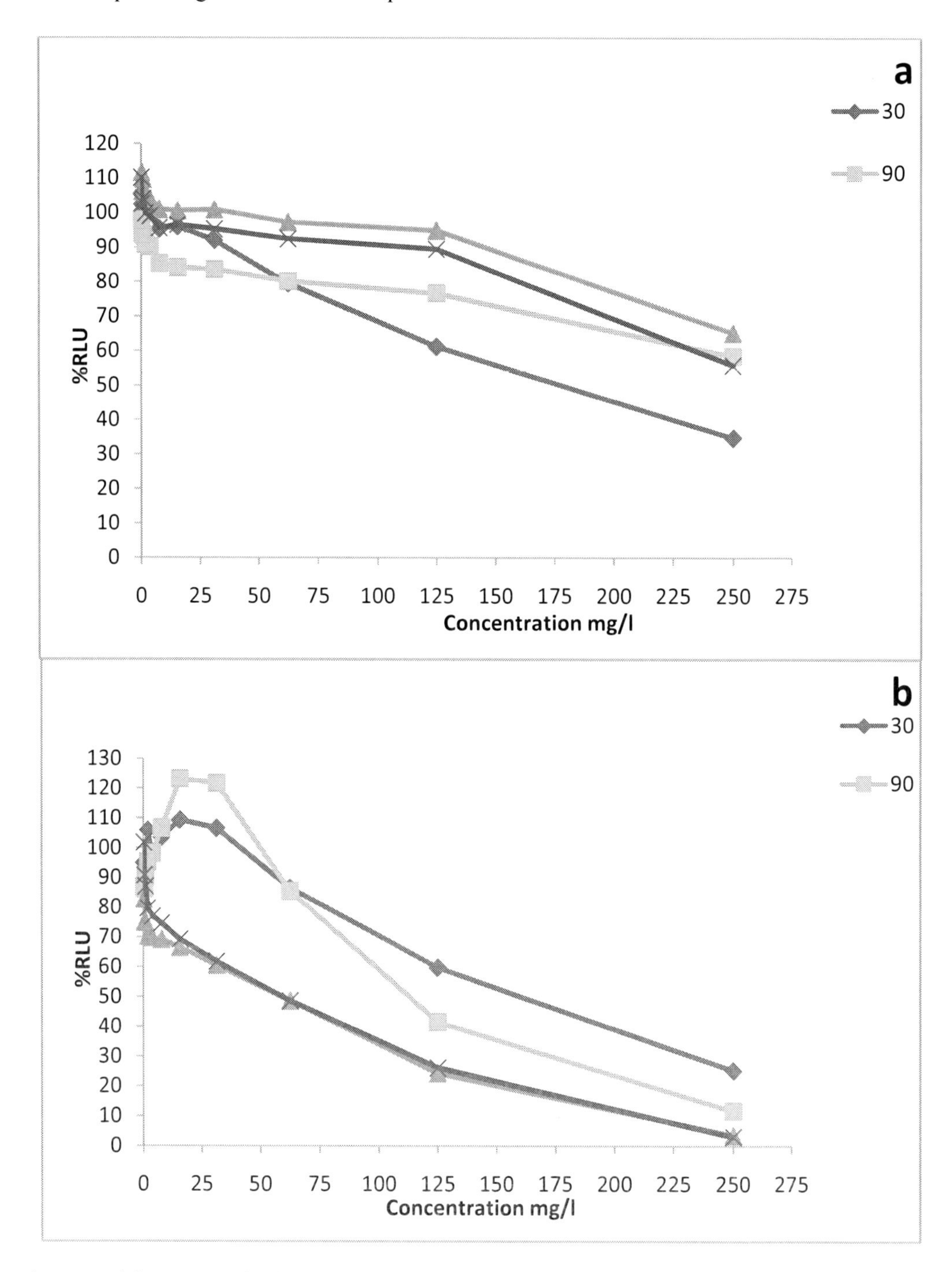

Figure 5. Light output reduction by P. putida BS566: luxCDABE when challenged with Hidroquinone (a) and chlorocathecol (b) after 30, 90, 80 and 240 min. Mean of three replicates with fifferent batches. Coefficient of variation 8 % among independent assays.

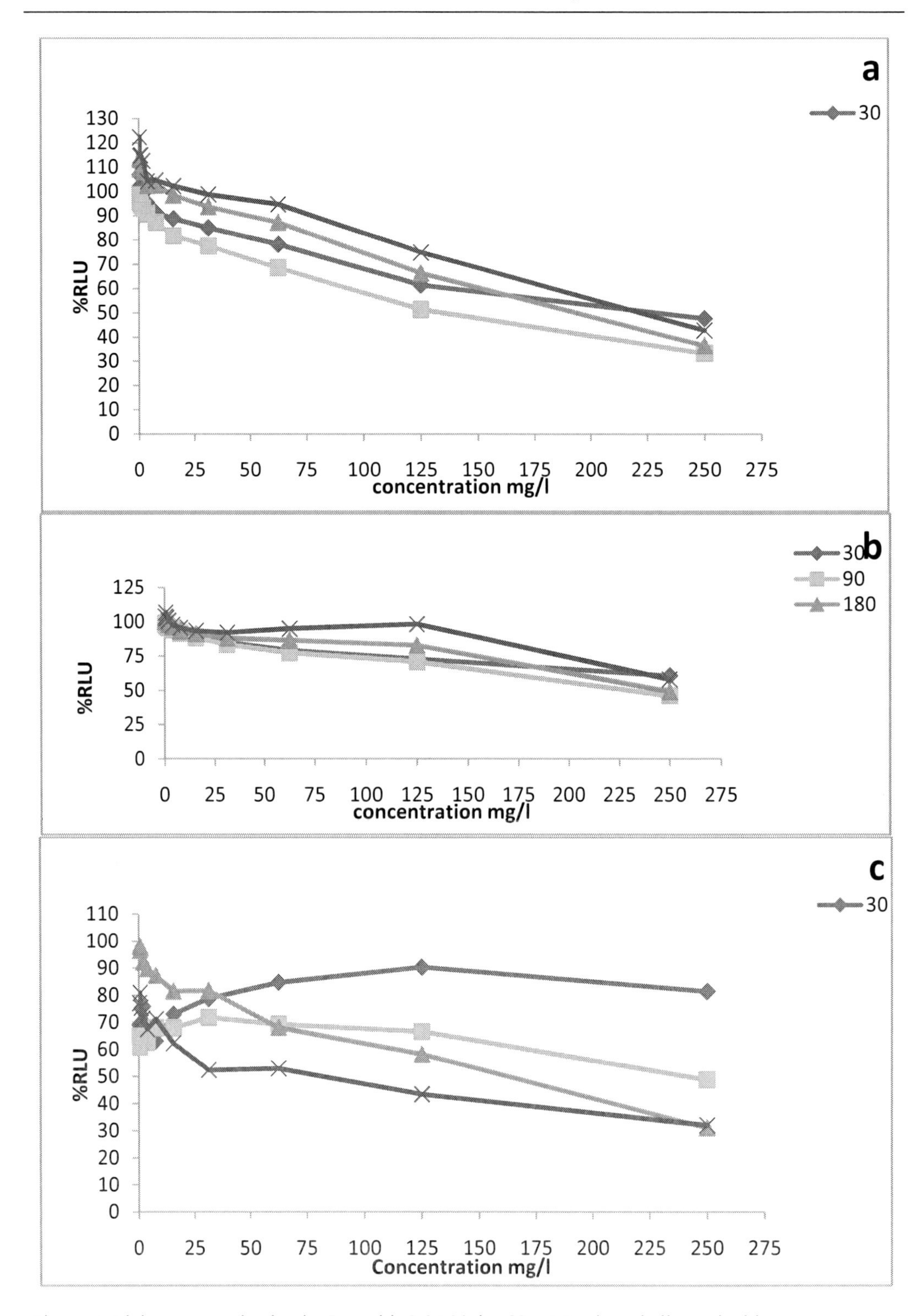

Figure 6. Light output reduction by P. putida BS566::luxCDABE when challenged with Chlororesorcinol (a), cathecol (b) and resorcinol (c) after 30, 90, 80 and 240 min. Mean of three replicates with fifferent batches. Coefficient of variation 8 % among independent assays.

Thus, understanding the fate and transport of phenolic compounds and transformation products in removal treatment plants is relevant in evaluating the impact of discharge in the environment and in developing a program for pollution control.

Silver Materials

IC_{50} values following challenge with $AgNO_3$, Ag-NP and Ag-MP are shown in Table 3. Calculated coefficients of variation (CV) between independent assays were found to be between 1% and 15%. It is well known that silver ion and silver based compounds are highly toxic to microorganisms and showing strong biocide effects to many bacteria species (Zhao and Stevens, 1998; Sondi and Salopek-Sondi, 2004; Brayner et al., 2006; Pal et al., 2007; Choi et al., 2008; Jiang et al., 2009).

Figure 7 shows the light output reduction by P. putida BS566::luxCDABE when challenged with 2500μg L-1 Silver ion. Among the silver species tested, AgNO3 is by far the most toxic to P. putida in this study highlighting the action of ionic Ag+ ($P<0.05$).

Figures 8-9 show the light output reduction by P. putida BS566::luxCDABE when challenged with Ag nanoparticles and Ag microparticles. When using BSA as stabilizer, Ag nanoparticles showed to be statistically more toxic than Ag microparticles ($P<0.05$). The same was observed when using citric acid as stabilizer.

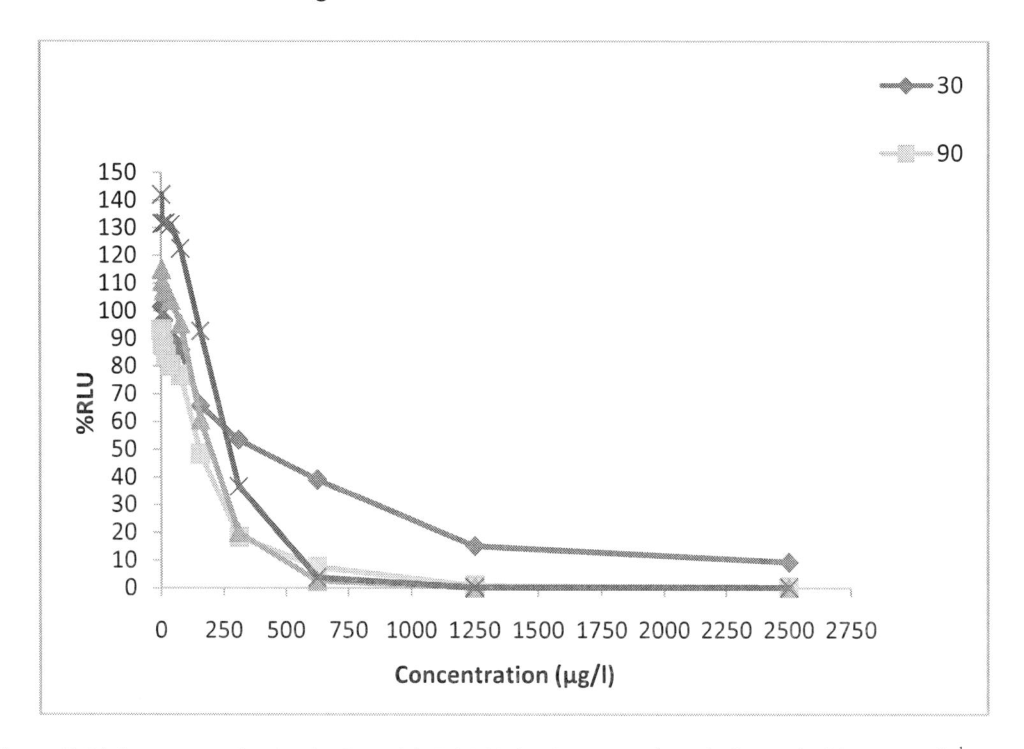

Figure 7. Light output reduction by P. putida BS566::luxCDABE when challenged with 2500μg L^{-1} Silver ion after 30, 90, 80 and 240 min. Mean of three replicates with different batches. Coefficient of variation 15% among independent assays.

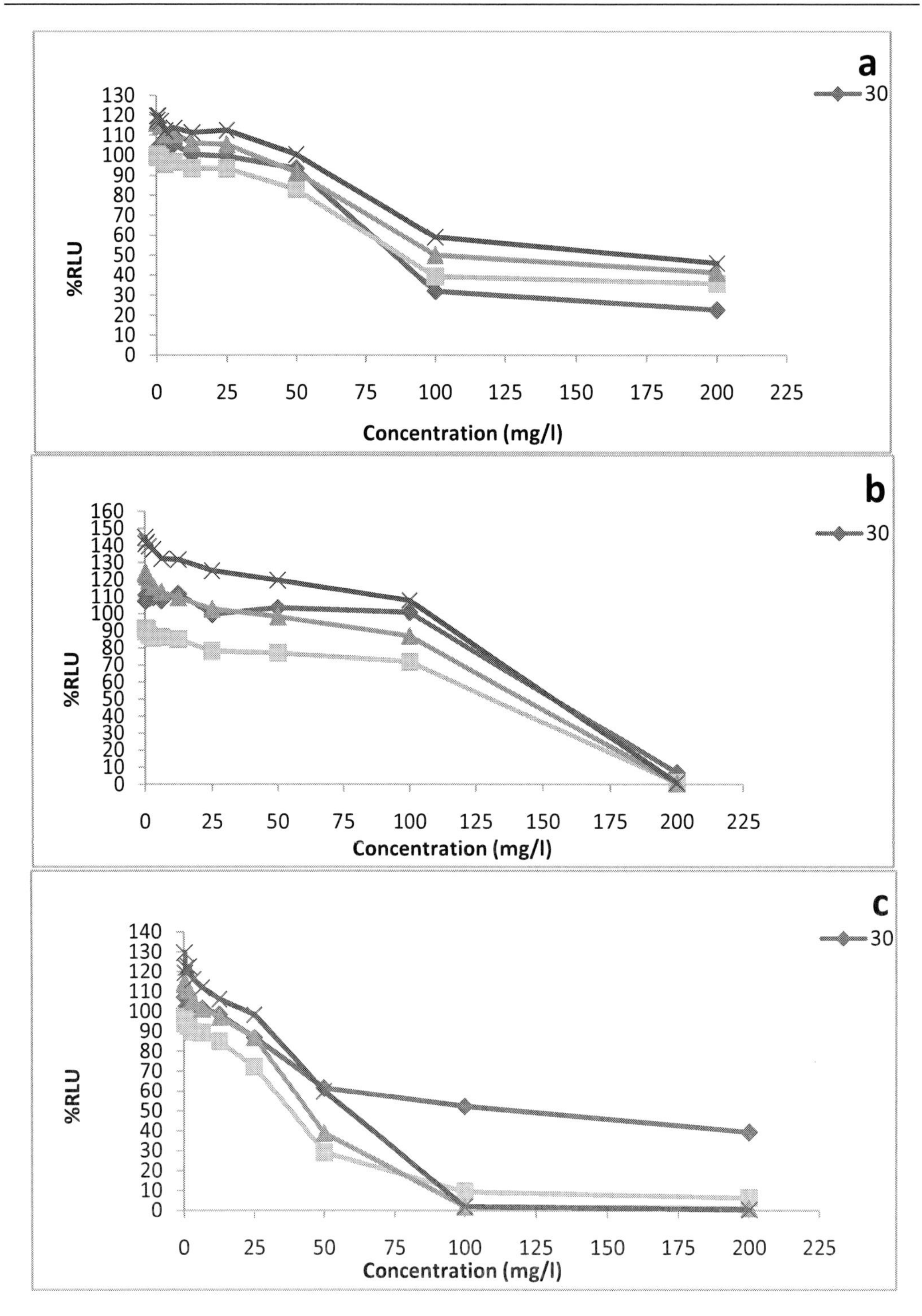

Figure 8. Light output reduction by P. putida BS566::luxCDABE when challenged with 200mg L-1 Ag nanoparticles (a) without stabilizer), (b) with 0.1% Citric acid and (c) with BSA after 30, 90, 80 and 240 min. Mean of three replicates with different batches. Coefficient of variation between 2 and 13% among independent assays.

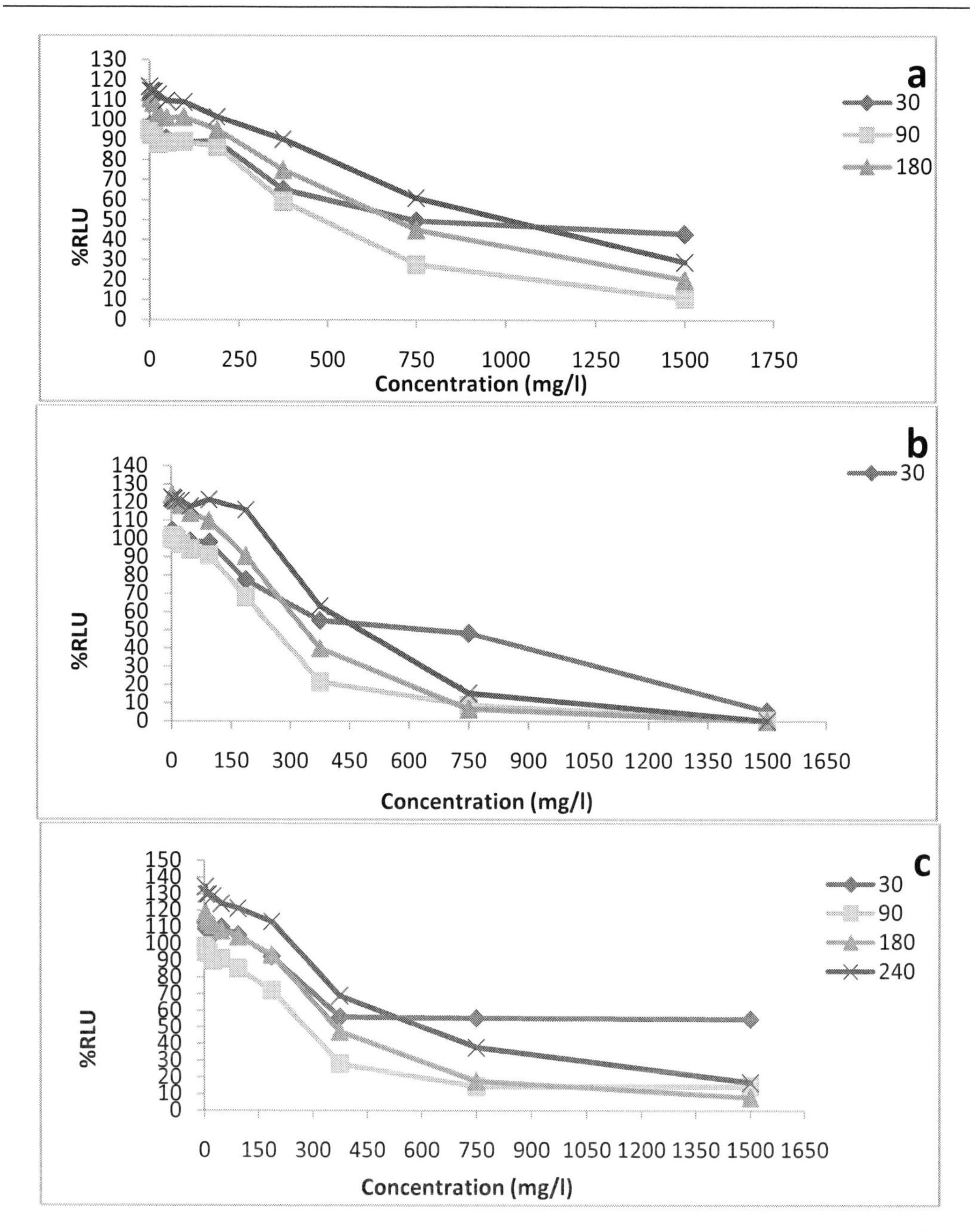

Figure 9. Light output reduction by P. putida BS566::luxCDABE when challenged with 1500mg L^{-1} Ag micro particles (a) without stabilizer), (b) with 0.1% Citric acid and (c) with BSA after 30, 90, 80 and 240 min. Mean of three replicates with different batches. Coefficient of variation between 4 and 14 % among independent assays.

Highest toxicity (Table 3) was obtained with the assay after 90 min incubation and indeed this was the case for all silver species tested with or without stabilisation. The order in toxicity was Ag+>Ag-NP (35 nm)>Ag-MP (0.6-1.6 μm). If we consider the 90 min assay results, that show the highest toxicity, to compare the data, it is seen that Ag-NP are nearly 200 times less toxic that Ag+, with Ag-MP ~3000 times less toxic in non-stabilised systems

($P<0.05$). The effect with Ag-NP was more pronounced in BSA systems with a calculated IC50 value ~4 times statistically lower than Ag-NP with citric acid ($P<0.05$). Both stabilisers appear to perform, increasing toxicity perhaps through better dispersion of the NP providing more surface area for Ag to have an effect. This however may not be the case as citric acid may be just as effective in preventing agglomeration. Citric acid is an effective chelating/complexing agent for metals in solution and perhaps its effect on the toxicity value is due to this fact acting to immobilise Ag^+ released in solution or those on surfaces. The use of different stabilisers and their effect on toxicity values must be assessed particularly if laboratory bioassay results will have significance in wastewater discharge consents.

There was no significant difference in the IC50 value obtained for Ag-MP with BSA and citric acid stabilisation ($P>0.05$) and both values were ~2 times statistically lower than Ag-MP tested without stabilisation ($P<0.05$).

The antimicrobial effect is related to the amount and the rate of silver released by NP. Is has been proposed that ionised silver binds to cell membrane proteins and severe structural changes occur in the bacterial cell wall leading to cell distortion and death (Lansdown et al., 2002; Castellano et al., 2007).

Silver is classified as the "soft" metal group (Blaser et al., 2008) and it complexes with many organic or inorganic materials such as chloride, sulphide, thiosulphate (Ratte, 1999). Thus, understanding the fate and transport of silver in waste water treatment plants is relevant in evaluating the impact of silver discharge in the environment and in developing a program for silver pollution control. Wastewater treatment plants are the final step to control silver discharge and our results demonstrate that the use of a bacterial biosensor as P. putida BS566::luxCDABE, native from a highly polluted environment, provides a robust, early warning system of acute toxicity which could lead to process failure. This strain is suitable for toxicity monitoring in a highly polluted industrial waste water treatment streams, including the ones with the presence of metals as silver. Silver NPs have become one of the nanomaterials most used in consumer product (Maynard and Michelson, 2006), it is considered as the most prevalent of engineered materials (Rejeski and Lekas, 2008) and is likely that nanowaste should increase and therefore enter the wastewater treatment plants. Blaser et al., (2008) pointed out that silver released to wastewater is incorporated into sewage sludge and may spread further on agricultural fields where will mainly stay in the top layer of soils (Hou et al., 2005). Landfilling of sewage sludge may allow silver to leach into subsoil and groundwater. Estimation of silver load in sewage sludge and its microorganisms growth inhibition has been predicted. Blaser et al., (2008) predicted that an expected silver concentration in sewage treatment plant range from 2 $\mu g\ L^{-1}$ to 18 $\mu g\ L^{-1}$. Shafer et al., (1998) reported a range of ~ 2 to 4 $\mu g\ L^{-1}$ of silver in sewage treatment plants treating common wastewater and a much higher load from industrial discharges (from 24 to 105 μg L-1).

The removal of silver ion by chloride free sludge is dependent on the silver-sludge loading, the solution pH and the concentration of dissolved organic matter. Studying the interactions of silver with wastewater constituents, Wang et al., (2003) showed that silver ion, under field conditions, can be removed through precipitation with chloride and adsorption by sludge particulates which increased in acidic conditions but decreased in alkaline conditions. However, the authors (Wang et al., 2003) pointed out that the formation of silver-ion-dissolved organic matter complexes, which is increased in alkaline conditions, reduces the silver ion adsorption by sludge.

Table 3. IC_{50} (mgL^{-1}) values for light emission reduction by P. putida BS566::luxCDABE after 30, 90, 180 and 240 min

Time (min)	30	90	180	240
$AgNO_3$	0.44	0.18	0.25	0.30
Ag NP	88	81	91.5	184
Ag NP –BSA	102	35	43	50
Ag NP –CA	147	126	136	149
Ag MP	715	530	765	1075
Ag MP – BSA	375	256	308	330
Ag MP- CA	700	240	300	337

Values presented are an average of at least 3 independent experiments carried out with different batches, standard deviation between 1 and 15%.

The information regarding the inhibition of microbial growth by different Ag species, especially in wastewater treatment systems, is valuable for operating planning and control. The presence and activity of microorganisms in biological wastewater treatment are vital to the process. One of the projected applications of such strains is its combined use as analytical panel for toxicant detection.

CONCLUSION

An important advantage of using this organism is that a positive response will not only indicate the presence of a toxicant but will also provide some idea as to its character.

In conclusion, our results demonstrate that the use of a bacterial biosensor as P. putida BS566::luxCDABE, native from a highly polluted environment, provides a robust, early warning system of acute toxicity which could lead to process failure. This strain is suitable for toxicity monitoring in a highly polluted industrial waste water treatment streams and it has the potential to inform on upstream process changes prior to their impacting upon the biological remediation section. Furthermore, as this reporter organism is based upon microorganisms naturally resident within the system, there remains the potential for its in situ deployment within the system with minimal effects upon their physiology.

ACKNOWLEDGMENTS

This work was part of a project funded by the Brazilian National Research and Development Council (CNPq). Also we would like to acknowledge the Environmental Research Laboratory at Edinburgh Napier University, Edinburgh, Scotland.

REFERENCES

Arnold S.M., Hickey W.J., Harris R.F. (1995). Degradation of atrazine by Fenton's reagent: Condition optimization and product quantification. *Environ. Sci. Technol.* 29, 2083-2089.

Aruoja, V., Dubourguiera, H-C., Kasemetsa, K. , Kahrua, A., 2009. Toxicity of nanoparticles of CuO, ZnO and TiO2 to microalgae *Pseudokirchneriella subcapitata. Sci. Total Environ.* 407, 1461-1468.

Auffan, M., Achouak, W., Rose, J., Roncato, R-A., Chaneac, C., Waite, D.T., Maison, A., Oicik, J.C., Wiesner, M.R., Vesbottero, J-Y. (2008). Relation between the Redox State of Iron-Based Nanoparticles and Their Cytotoxicity toward *Escherichia coli. Environ. Sci. Technol. 42, 6730–6735.*

Belkin,S., Smulski, D.R., Vollmer,A.C., Van Dyk, T.K., Larossa, R.A. (1996). Oxidative Stress Detection with *Escherichia coli. Appl. Environ.Microbiol.*62, 2252–2256.

Belkin, S., Smulski D.R., Dadon S., Vollmer, A.C., Van Dyk, T.K., Larossa, R.A. (1997). Panel of stress-responsive luminous bacteria for the detection of selected classes of toxicants. *Water Res.* 31,3009-3016.

Belkin, S. (2007). Genetically engineered microorganisms for pollution monitoring. In: Soil and Water Pollution Monitoring, Protection and Remediation, NATO Science Series: IV: *Earth and Environmental Sciences*, Volume 69, 147-160.

Ben-Israel, O., Ben-Israel, H., Ulitzur,S. (1998). Identification and Quantification of Toxic Chemicals by Use of *Escherichia coli* Carrying *lux* Genes Fused to Stress Promoters. *Appl. Environ.Microbiol.* 64, 4346-4352.

Benitez F.J., Acero J.L., Real F.J. and Garcia J. (2003). Kinetics of photodegradation and ozonation of pentachlorophenol. *Chemosphere* 51, 651–662.

Blaser SA, Scheringer M, MacLeod M, Hungerbuhler K (2008). Estimation of cumulative aquatic exposure and risk due to silver: contribution of nano-functionalised plastics and textiles. *Sci. Total Environ. 390*, 396-409.

Blinova, I., Ivaska, A., Heinlaan, M., Mortimer, M., Kahru, A. (2010). Ecotoxicity of nanoparticles of CuO and ZnO in natural water. *Environ. Pollut.* 158, 41–47.

Boyd E., KIllham K., Mcharg A.A. (2001). Toxicity of mono-, di- and tri-chlorophenols to *lux* marked terrestrial bacteria, *Burkholderia* species *Rasc* c2 and *Pseudomonas fluorencens*. *Chemospher*e, 43, 157-166.

Brayner, R., Ferrari-Iliou, R., Brivois, N., Djediat, S., Benedetti, M.F., Fie′vet, F. (2006). Toxicological Impact Studies Based on *Escherichia coli* Bacteria in Ultrafine ZnO Nanoparticles Colloidal Medium. *Nanno Lett.* 866-870.

Brown, J.S., Rattray, E.A.S., Paton, G.I., Reid, G., Gaffor, I., Killham K. (1996). Comparative assessment of the toxicity of a papermil effluent by respiratory and luminescence based bacterial assay. *Chemosphere* 32, 1553-1561.

Bunce, N., Liu, L., Zhu, J.,Lane, D. (1997). Reaction of naphthalene and its derivatives with hydroxyl radicals in the gas phase. *Environ, Sci.Technol.* 31, 2252–2259.

Bystrzejewska-Piotrowska, G., Golimowski, J., Urban. P.L. (2009). Nanoparticles: Their potential toxicity, waste and environmental management. *Waste Manage.* 29, 2587–2595.

Castellano, J.J., Shafii, S.M., Ko, F., Donate, G., Wright, T.E., Mannari, R.(2007). Comparative evaluation of silver-containing antimicrobial dressings and drugs. *International Wound Journal,* 4, 114 – 122.

Chi, J., Huang, G. (2002). A multimedia fugacity river model of pentachlorophenol in south drainage canal, China. *J. Environ. Sci..Health A* 37, 113–25.

Chen, R., Pignatello, J. (1997). Role of quinone intermediates as electron shuttles in Fenton and photoassisted Fenton oxidation of aromatic compounds. *Environ. Sci.Technol.* 31, 2399–2406.

Choi, O., Deng, K.K., Kim, N-J., Ross Jr,, L, Surampalli, R.Y., Hu, Z. (2008). The inhibitory effects of silver nanoparticles, silver ions, and silver chloride colloids on microbial growth. *Water Res*. 42, 3066-3074.

Choi, S.H., Gu, M.B. (2003). Toxicity biomonitoring of degradation byproducts using freeze-dried recombinant bioluminescent bacteria. *Chimica Acta* 481, 229–238.

Chu, C.L., Lin,P.H., Dong, Y.S., Guo, D.Y. (2002). Influences of citric acid as a chelating reagent on the characteristics of nanophase hydroxyapatite powders fabricated by a sol-gel method. *J. Mat. Sci. Lett.* 21, 1793-1795.

Copley, S.D. (2000). Evolution of a metabolic pathway for degradation of a toxic xenobiotic: the patchwork approach. *Trends Biochem.Sci.*25, 261-265.

Cronin, M.T.D., Schultz, T.W.(1998). Structure–Toxicity Relationships for Three Mechanisms of Action of Toxicity to *Vibrio fischeri*. (1998). *Ecotoxicol. Environ. Saf.* 39, 65-69.

Damianovic, M.H.R.Z., Moraes, E.M., Zaiat, M., Foresti E. (2009). Pentachlorophenol (PCP) dechlorination in horizontal-flow anaerobic immobilized biomass (HAIB) reactors. *Bioresource Technol.*100, 4361–4367.

Dennison, M.J., Turner, A.P.F. (1995). Biosensors for environmental monitoring. *Biotechnol. Adv.* 13, 1-12.

Eker, S., Kargi, F. (2007). Performance of a hybrid-loop bioreactor system in biological treatment of 2,4,6-trichlorophenol containing synthetic wastewater: effects of hydraulic residence time. *J. Hazard. Mater.* 144, 86–92.

Escher, B.I., Snozzi, M., Schwarzenbach, R.P. (1996). Uptake, speciation and uncoupling activity of substituted phenols in energy transducing membranes. *Environ. Sci. Technol.* 30, 3071-3079.

Fallmann, H., Krutzler, R., Bauer, S., Blanco, M.J. (1999). Applicability of the Photo-Fenton method for treating water containing pesticides. *Catal. Today* 54, 309-319.

Foucaud, L., Wilson, M.R., Brown, D.M.,Stone ,V. (2007). Measurements of reactive species production by nanoparticles prepared in biologically relevant media. *Toxicol. Lett.* 174, 1-9.

Fukushima, M., Kawasaki, M., Sawada, A., Ichikawa, H., Morimoto, K., Tatsumi, K., Tanaka, S. (2002). Facilitation of pentachlorophenol degradation by the addition of ascorbic acid to aqueous mixtures of tetrakis(sulfonatophenyl) porphyrin and iron(II) *J. Mol. Catal A- Chem* 187, 201–213.

Gomez, M., Murcia,M.D., Gomez, E., Gomez, J.L., Dams, R., Christofi, N. (2010). Enhancement of 4-Chlorophenol Photodegradation with KrCl Excimer UV Lamp by Adding Hydrogen Peroxide. *Separ. Sci. Technol.* 45, 1603–1609.

Gorska, P., Zaleska, A., Hupka, J. (2009). Photodegradation of phenol by UV/TiO2 and Vis/N,C-TiO2 processes: Comparative mechanistic and kinetic studies. *Sep. Purif. Technol.* 68, 90–96.

Gu, M.B., Choi, S.H. (2001). Monitoring and classification of toxicity using recombinant bioluminescent bacteria. *Water Sci. Technol.* 43, 147–154.

Hirvonen, A., Trapido, M., Hentunen, J., Tarhanen, J. (2000). Formation of hydroxylated and dimeric intermediates during oxidation of chlorinated phenols in aqueous solutions. *Chemosphere* 41, 1211–1218.

Heinlaan, M., Ivask, A., Blinova, I., Dubourguier, H-C., Kahru, A. (2008). Toxicity of nanosized and bulk ZnO, CuO and TiO2 to bacteria Vibrio fischeri and crustaceans Daphnia magna and Thamnocephalus platyurus. *Chemosphere,* 71, 1308-1316.

Hollender, J., Hopp, J., Dott, W. (1997). Degradation of 4-Chlorophenol via the *meta* Cleavage Pathway by *Comamonas testosteroni* JH5. *Appl. Environ. Microbiol.* 63, 4567–4572.

Hong A.P.K., Zeng Y. (2002). Degradation of pentachlorophenol by ozonation and biodegradability of intermediates. *Water Res.* 36, 4243–4254.

Hou, H., Takamatsu, T., Koshikawa, M.K., Hosomi, M. (2005). Migration of silver, indium, tin, antimony and bismuth and variations in their chemical fractions on addition to uncontaminated soils. *Soil Sci.* 170, 624-639.

Jiang, W., Mashayekhi, H., Xing, B. (2009). Bacterial toxicity comparison between nano- and micro-scaled oxide particles. *Environ. Poll.* 157,1619–1625.

Kang, N., Lee, D.S., Yoon, J. (2002). Kinetic modelling of Fenton oxidation of phenol and monochlorophenols. *Chemosphere,* 47, 915–924.

Keith, L.H. , Telliard, W.A. (1979). Priority pollutants: a prospective view. *Environ. Sci. Technol.* 13,416–424.

Kim, J., Kuk, E., Yu, K., Kim, J., Park, S., Lee, H., Kim, S., Park, Y., Park, C., Wang, C. (2007). Antimicrobial effects of silver nanoparticles. *Nanomedicine: Nanotechnology, Biology and Medicine,* 3, 95-101.

Landsdown, A.B.C.(2002). Silver I: its antibacterial properties and mechanism of action. *J. Wound Care*, 11:125-138.

Leppard, G.G., Mavrocordatos, D., Perret, D. (2003).Electron-optical characterization of nano- and micro-particles in raw and treated waters: an overview. In: Boller M, editor. Proceedings of nano and microparticles in water and wastewater treatment, 50 (12). *Water Sci Technol,* 1–8.

Litchfield, C.D., Rao, M. (1998). Pentachlorophenol biodegradation:laboratory and field studies. In: *Biological treatment of hazardous wastes.* (L.J. DeFilippi, Ed). Wiley, New York, pp. 271–283.

Luo, T., Ai Z., Zhang, L. (2008). Fe@Fe2O3 core-shell nanowires as iron reagent 4 Sono-Fenton degradation of pentachlorophenol and the mechanism analysis. *Journal of Physics and Chemistry* 112, 8675–8681.

Madsen, T., Aamand, J. (1990). Effects of sulforoxy anions on degradation of pentachlorophenol by a methanogenic enrichment culture. *Appl. Environ. Microbiol.* 57, 2453–2458.

Maynard, A.D., Michelson, E. (2006). *The Nanotechnology Consumer Product Inventory* (http://nanotechnproject.org/44).

McAllister, K.A., Lee, H., Trevors, J.T. (1996). *Microbial degradation of pentachlorophenol. Biodegradation.* 7, 1–40.

Miller, R.M., Singer, G.M., Rosen, J.D., Bartha, R. (1988). Sequential degradation of chlorophenols by photolytic and microbial treatment. *Environ. Sci. Technol.* 22, 1215 - 1219.

Mills, G., Hoffmann, M.R. (1993). Photocatalytic degradation of pentachlorophenol on TiO2 particles: Identification of intermediates and mechanism of reaction. *Environ. Sci.Technol.* 27, 1681-1689.

Mortimer, M., Kasemets, K., Kurvet, I., Heinlaan, M., Kahru, A. (2008). Kinetic Vibrio fischeri bioluminescence inhibition assay for study of toxic effects of nanoparticles and colored/turbid samples. *Toxicol. in Vitro*, 22, 1412-1417.

Pal, S., Tak, Y.K., Song, J.M. (2007). Does the antibacterial activity of silver nanoparticles depend upon the shape of the nanoparticle? A study of the Gram-negative bacterium *Escherichia coli. Appl. Environ. Microbiol.* 73, 1712–1720.

Paton, G.I., Campbell, C.D., Glover, L.A., Killham, K. (1995). Assessment of bioavailability of heavy metals using lux modified constructs of *Pseudomonas fluorescens. Lett. Appl. Microbiol.* 20, 52–56.

Paton, G.I., Palmer, G., Burton, M., Rattray, E.A., McGrath, S.P., Killham K. (1997). Development of an acute and chronic ecotoxicity assay using *lux*-marked *Rhizobium leguminosarum* biovar *trifoli. Lett.Appl.Microbiol.* 24, 296-300.

Paulus, W. (1993). *Microbiocides for the protection of materials*.London, Chapman & Hall, London, United Kingdom, 526pp.

Pera-Titus, M., Garcia-Molina, V., Ban′ os, M., Gimenez, J., Esplugas, S. (2004). Degradation of chlorophenols by means of advanced oxidation processes: a general review. *Appl.Catal B- Environ.* 47, 219–256.

Ratte, H.T. (1999). Bioaccumulation and toxicity of silver compounds: a review. *Environ. Toxicol. Chem.* 18, 89-108.

Rejeski, D., Lekas, D. (2008). Nanotechnology field observations: scouting the new industrial waste. *J. Cleaner Prod.* 16, 1014-1017.

Shafer, M.M., Overdier, J.T., Armstong, D.H. (1998). Removal portioning and fate of silver and other metals in wastewater treatment plants and effluent-receiving streams. Environ. *Toxicol. Chem.* 17(4), 630-641.

Siejak, P., Frackowiak, D. (2007). Interactions between colloidal silver and photosynthetic pigments located in cyanobacteria fragments and in solution. *J. Photochem.Photobio B.* 88, 126-130.

Sinha, R., Karan, R., Sinha, A., Khare, S.K. (in press). Interaction and nanotoxic effect of ZnO and Ag nanoparticles on mesophilic and halophilic bacterial cells. *Bioresour. Technol.* (2010), doi:10.1016/j.biortech.2010.07.117.

Slabbert, J.L. (1986). Improved bacterial growth test for rapid water toxicity screening. *Bull. Environ. Contam. Toxicol.* 37, 565–569.

Sondi, I., Salopek-Sondi,B. (2004). Silver nanoparticles as antimicrobial agent: a case study on *E. coli* as a model for Gram-negative bacteria. *J.Colloid. Interf. Sci.* 275, 177-182.

Stefan, M.I., Bolton, J.R. (1998). Mechanism of the degradation of 1,4-dioxane in dilute aqueous solution using the UV/ hydrogen peroxide process. *Environ. Sci. Technol.* 32, 1588-1595.

Sundstrom, D.W., Weir, B.A., Klei, H.E. (1989). Destruction of aromatic pollutants by UV light catalyzed oxidation with hydrogen peroxide. *Environ. Prog.* 8, 6-11.

Wang, J., Huan, C.P., Pirestan, D. (2003). Interactions of silver with wastewater constituents. *Water Res.* 37, 4444-4452.

Weavers, L., Malsmatdt, N., Hoffman, M. (2000). Kinetics and mechanism of pentachlorophenol by sonication, ozonation, and sonolytics ozonation. *Environ. Sci. Technol.* 34, 1280–1285.

Westerhoff, P., Zhang, Y., Crittenden, J., Chen, Y.(2008). Properties of commercial nanoparticles that affect their removal during water treatment. In: Grassian VH, editor. Nanoscience and Nanotechnology*: Environmental and Health Impacts.* NJ: John Wiley and Sons, 71–90.

Whiteley, A.S., Wiles, S., Lilley, A.K., Philp, J., Bailey, M.J. (2001). Ecological and physiological analyses of Pseudomonad species within a phenol remediation system. *J. Microbiol. Meth.* 44, 79– 88.

Wiles, S., Whiteley, A.S., Philp, J.C., Bailey, M.J. (2003). Development of bespoke bioluminescent reporters with the potential for in situ deployment within a phenolic-remediating wastewater treatment system. *J. Microbiol. Meth.* 55, 667– 677.

Winson, M., Swift, S., Hill, P.J., Sims, C.M., Griesmayr, G., Bycroft, B.W., Williams, P., Gordon, S.A.B., Stewart, G.S.A.B. (1998). Engineering the luxCDABE genes from *Photorhabdus luminescens* to provide a bioluminescent reporter for constitutive and promoter probe plasmids and mini-Tn5 constructs. *FEMS Microbiol. Lett.* 163,193-202.

Wu, W.M., Hickey, R.F., Bhatnagar, M.K., Zeikus, J.C. (1989). Fatty acid degradation as a toll to monitor anaerobic sludge activity and toxicity. In: *44th Purdue University Industrial Waste Conference Proceedings*, pp. 225–233.

Zeng, Y., Wackett, L. (2008). Pentachlorophenol Family Pathway Map. http://umbbd.ethz.ch/pcp/pcp_map.html. 16/06/2010.

Zhao, G., Stevens, E. (1998). Multiple parameters for the comprehensive evaluation of the susceptibility of *Escherichia coli* to the silver ion. *BioMetals*, 11, 27-32.

In: Bioluminescence
Editor: David J. Rodgerson, pp. 97-114
ISBN 978-1-61209-747-3

Chapter 5

BIOLUMINESCENCE: CHARACTERISTICS, ADAPTATIONS AND BIOTECHNOLOGY

Iram Liaqat
School of Biomedical, Biomolecular & Chemical Sciences,
Microbiology M502,
The University of Western Australia,
35 Stirling Hwy Crawley WA 6009, Australia

ABSTRACT

Bioluminescence is a biological phenomenon in which energy is released by a chemical reaction in the form of cold light emission (chemiluminescence). Evolution of bioluminescence has arisen independently as many as 30 times with the five main traits of camouflage, attraction, repulsion, communication and illumination. Bioluminescence is common in bacteria and fungi but is also observed in ctenophores, annelid worms, jelly fish, mollusks, glow worms and fireflies. Bioluminescent symbiotic microorganisms are frequently found in symbiotic association with larger organisms, such as fish. Bioluminescence is caused by pigments produced by the organisms, such as luciferin, coelenterazine, and vargulin. The synthesis of luciferin, the most studied bioluminescent pigment, is catalyzed by the enzyme luciferase that is bound in inactive form to adenosine triphosphate (ATP). Once released from ATP, luciferase causes luciferin to react with molecular oxygen, yielding an electronically excited oxyluciferin species. Relaxation of excited oxyluciferin to its ground state results in emission of visible light. Slight structural differences in luciferase affect both the high quantum yield of the luciferin/luciferase reaction and bioluminescence colour. The luciferin and catalyzing enzyme, as well as a co-factor such as oxygen, are bound together to form a single unit called a "photoprotein". A photoprotein can be triggered to produce light when a particular type of ion (frequently calcium) is added. Bioluminescent pigments can be expressed in recombinant organisms and found numerous applications in biological, medical, and biotechnology research. Luciferases are more sensitive than fluorescent reporters, owing to the extremely low background levels of bioluminescence. Also, there is no need for exogenous illumination, which in fluorescence methods can bleach the reporter, perturbs physiology in light-sensitive tissues (e.g. the retina), and causes phototoxic damage to cells. Therefore, bioluminescence imaging is used to study viable cells, tissues and whole organisms. Molecular imaging techniques represent a revolutionary advancement in our ability to study structural and functional relationships in biology by combining the disciplines of molecular/ cellular biology and imaging technology. Newly developed imaging methods allow transcriptional/translational regulation, signal transduction, protein-protein interaction, oncogenic transformation, cell and protein trafficking, and target drug action to be monitored in vivo in real-time with high temporal and spatial resolution, thus providing researchers with priceless information on cellular functions. Those bioluminescent pigments that emit in the visible range can be used for high-throughput screening methods, which are at the forefront of renewable energy applications. Recently, even eukaryotic parasites Plasmodium, Leishmania and Toxoplasma have been transformed with luciferase and yielded unique insights into their in vivo behavior. The applications to biosensing are also interesting and include environmental monitoring of toxic and mutagenic compounds, heavy metals, etc. Other possible applications of engineered bioluminescence include engineering of "smart" organisms and devices, such as glowing trees on highways, novelty bioluminescent pets, bioluminescent agricultural crops and plants at time of watering and bio-identifiers for escaped convicts or mentally ill patients. The discovery of other bioluminescent pigments that emit at different wavelength and the advancements in recombinant technology will in future increase the number of applications.

Keywords: bioluminescence, adaptation, characteristics, biotechnology perspective

Introduction

Bioluminescence, a form of chemiluminescence, the emission of visible light by an organism as a result of a natural chemical reaction [Haddock et al., 2010] has intrigued humankind from the earliest history. The first evidence to a luminous animal was found in the thirteen classics of China. Afterwards, great diversity of organisms were discovered with the ability to emit and display light controlling it in a variety of ways. Bioluminescence has evolved independently as many as 30 times; thus the responsible genes are unrelated in bacteria, unicellular algae, coelenterates, beetles, fishes, and others. However, there are several examples of cross-phyletic similarities among the substrates; Some of these may be accounted for nutritionally, but in other cases they may have evolved independently [Hastings, 1983].

However, bioluminescence shouldn't be confused with other natural optical phenomena such as fluorescence and phosphorescence. Fluorescent molecules don't produce their own light; they absorb photons, which temporarily excite electrons to a higher energy state. Relaxation of these electrons to ground state results in release of energy. This complete reaction occurs very rapidly (usually within picoseconds to microseconds). Hence, fluorescent light is only seen while the specimen is being illuminated. Examples are chlorophyll, phycobiliproteins, and the green fluorescent proteins (GFPs) [Haddock *et al.*, 2010].

During the past few years considerable progress has been made to understand mechanism of the bioluminescent reaction. The chemical events leading to an excited state are reasonably well established both for bioluminescence or chemiluminescence. The actual emission of chemiluminescence or bioluminescence (a chemiluminescence that requires an enzyme) is the extremely rapid process. It requires at least two different chemicals, generally called luciferin and luciferase, which end up in generation of a molecule in an electronically excited state, oxyluciferin [Haddock, 2001]. Luciferin is the general term used to refer to the chemical that actually produces the light, and reaction rate is controlled by an enzyme called as luciferase or photoproteins. In inactive form, luciferase is bound to adenosine triphosphate (ATP), when bioluminescence occurs, it is released from ATP, catalyses the oxidation of the luciferin producing excited oxyluciferin [Haddock *et al.*, 2010].

The excited state of the emitter has a very short lifetime. It holds the reaction energy for no more than a few nanoseconds before releasing it in the form of a photon (Figure-1). Slight structural differences in luciferase affect both the high quantum yield of the luciferin/luciferase reaction and bioluminescence colour. Photoproteins, a luciferase variant in which factors required for light emission (including the luciferin and oxygen) bound together as one unit, are triggered to produce light upon binding another ion or cofactor, such as Ca^{2+} or Mg^{2+}, which causes a conformational change in the protein. This gives the organism a way to accurately control light emission [Haddock *et al.*, 2010].

Of the two main components for bioluminescent reaction, the luciferins are highly conserved across phyla. Four luciferins are responsible for most light production in the ocean (Figure-2), although many other light-emitting reactions are yet need to be discovered. Despite the fact, that luciferins are conserved, luciferases and photoproteins are unique and derived from many evolutionary lineages. For example, all hydrozoans use photoproteins as luciferin, but all Cnidarians use coelenterazine; scyphozoans primarily luciferases; and anthozoans unrelated luciferases in conjunction with a luciferin-binding protein.

In other words, each luminescent hydrozoan species has one or more genes coding for a photoprotein, sequence of which can be readily aligned with other hydrozoan photoproteins, but this family of sequences shows little to no correspondence with the luciferases from other cnidarians. It is also interesting that chemically identical luciferin can be the active compound in unrelated organisms. Example include the coelenterazine, which is found in at least nine phyla, jellyfish, crustaceans, spanning protozoans, molluscs, vertebrates and arrow worms [Haddock *et al.*, 2010].

However this doesn't mean that same type of coelenterazine is found in all organisms. Different organisms are synthesizing it differently. In some it is acquired exogenously through diet [Haddock *et al.*, 2001] while other obtain it by some chemical reaction. Another explanation of different coelenterazine is that it is found in both luminous and non luminous organisms. However, research is required to understand the complete biosynthesis pathway of marine luciferins. Several factors affect the color of a bioluminescence.

In the simplest case, the emission matches the fluorescence of an excited luciferase-bound product of the reaction. The luciferase structure can itself alter the color, for example, single amino acid substitutions of luciferase in firefly result in significant shifts in the emission spectrum. Whereas, the chromophores of accessory proteins associated with a luciferase such as the yellow fluorescent protein (YFP) in bacteria and GFP in coelenterates may serve as alternate emitters [Haddock *et al.*, 2010].

The mechanism and regulation of bioluminescence differ among various groups. Bacteria and some other systems emit light continuously, and in many others, luminescence occurs as flashes, lasting for 0.1–1 s duration. So it requires a rapid turn on and off mechanism controlled in an enzymatic way, with reagents sequestered appropriately and subject to quick mobilization. In coelenterates, flashing is triggered by calcium entry, and the calcium sites on the relevant proteins have homologies with calmodulin, whereas in fireflies, Oxygen is main candidate with unknown triggering agent. In Dinoflagellate, scintillons are novel cytoplasmic structures whose flash is triggered by a rapid pH change [Wilson and Hastings, 1998].

A great diversity of organisms have the ability to produce and display light as well as to control it many different ways. Basic research, carried out initially for the pleasure of uncovering how organisms convert chemical energy into light, can now be credited with remarkable advances in entirely unrelated fields.

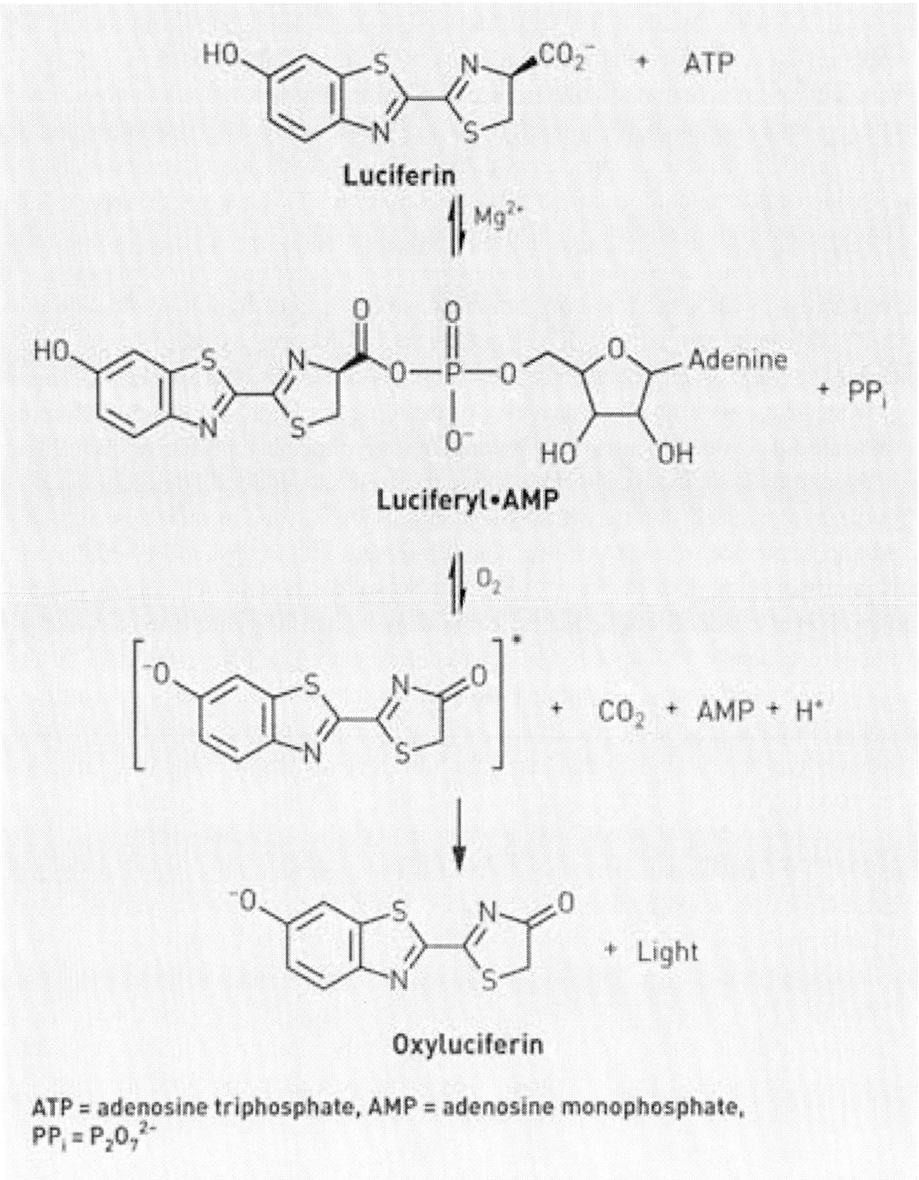
+ ATP
Luciferin
Mg2+
Adenine
+ PPi
Luciferyl•AMP
O2
+ CO2 + AMP + H+
+ Light
Oxyluciferin
ATP = adenosine triphosphate, AMP = adenosine monophosphate,
PPi = P2O7 2-

Figure 1. Mechanism of bioluminescence. Chemical & Engineering News, ISSN 0009-2347, Copyright © 2010 American Chemical Society.

An important example stems from the study of bacterial luminescence, where the discovery of autoinduction led to the notion of "quorum sensing," a chemical communication between bacteria, which is throwing light on processes ranging from pathogenesis to symbiosis (Liaqat *et al.*, 2010). We actually owe the discovery of a way of intercellular communication called as quorum sensing to bioluminescent bacteria. Isolated bacteria, free-living at sea, spend energy do not emit light. Since the relevant genes are not turned on. For example, in culture, *V. harveyi* and *V. fischeri* first grow without synthesizing luciferase until mid or late exponential phase is attained and a cell density threshold is crossed. At this stage transcription of specific genes is initiated, after which luciferase is expressed and the cells luminesce. Thus by sensing the level of autoinducer, the cells are able to estimate their density and to initiate the energetically costly synthesis of luciferase [Wilson and Hastings, 1998]. Another important example is GFP, discovered as accessory emitter in coelenterate bioluminescence and now used as a reporter protein [Wilson and Hastings, 1998]. Although there is a comprehensive list of bioluminescent organisms, however bacteria, dinoflagellates, ctenophores, cnidarians, annelids, worms, mollusks, crustaceans, ecinodersm and fish have been studied in detailed for research hence discussed below.

Bioluminescent bacteria are common in the ocean, particularly in temperate to warmer waters [Dunlap and Kita-Tsukamoto, 2006]. They may be cultured from almost any piece of detritus or tissue found on the beach, and even from uncooked seafood if known to glow after being left for a time. Bacteria are luminous only after they have reached sufficiently high concentrations to initiate quorum sensing [Nealson and Hastings, 2006], and once induced, they glow continuously. These properties are unique to bacteria, making them suitable as photogenic symbionts. Among prokaryotes, light production is limited to eubacteria only, specifically Gram-negative γ-proteobacteria. The best-studied symbiotic bacteria are in the genus *Vibrio*, including the predominantly free-living species *V. harveyi* (*Beneckea harveyi*) [Makemson *et al.*, 1997]. *Vibrio fischeri* (often called *Photobacterium*) is part of the species-complex involved in symbiosis with sepiolid and loliginid squid and monocentrid fishes, while *Photobacterium leiognathi* and relatives are primarily symbionts for leiognathid, apogonid, and morid fishes [Kaeding *et al.*, 2007]. Evolutionary dynamics of several strains make them suitable symbionts for light-organ colonization. Nonetheless, the best-studied mutualism between the bobtail squid *Euprymna scolopes* and *Vibrio fischeri*, showed that the presence of the bacteria actually induces the morphological development of the squid light organ [McFall-Ngai and Ruby, 1998].

While host squid also monitors the luminescent performance of the symbionts and strains that fail to maintain adequate light production are rejected by a yet unknown mechanism [Nyholm *et al.*, 2004]. A whole-genome sequence of *V. fischeri* provides robust support for the idea that the same mechanisms that allow disease-causing enteric Vibrionaceae to infect human hosts (*V. cholerae*, *V. parahaemolyticus*) may be at work in establishing beneficial symbioses with marine species [Ruby *et al.*, 2005]. In fact, one researcher is said to have infected himself for a period of months when working with luminous *Photobacterium leiognathi* [Campbell, 2008].

Next to bacteria, dinoflagellates are the most commonly studied bioluminescent organisms. There are at least 18 luminous genera [Baker *et al.*, 2008], including *Gonyaulax* (*Lingulodinium*), *Noctiluca*, *Protoperidinium*, and *Pyrocystis*. Dinoflagellates invest much in their effort to luminesce, and offer great energy to bioluminescence before growth, although luminescence comes after swimming ability [Latz and Jeong 1996]. They usually produce sparkling lights in the water seen by swimmers, sailors and beachgoers. They produce the "bioluminescent bays" which are tourist destinations in Puerto Rico and Jamaica. These protists can be autotrophic (photosynthetic) or heterotrophic.

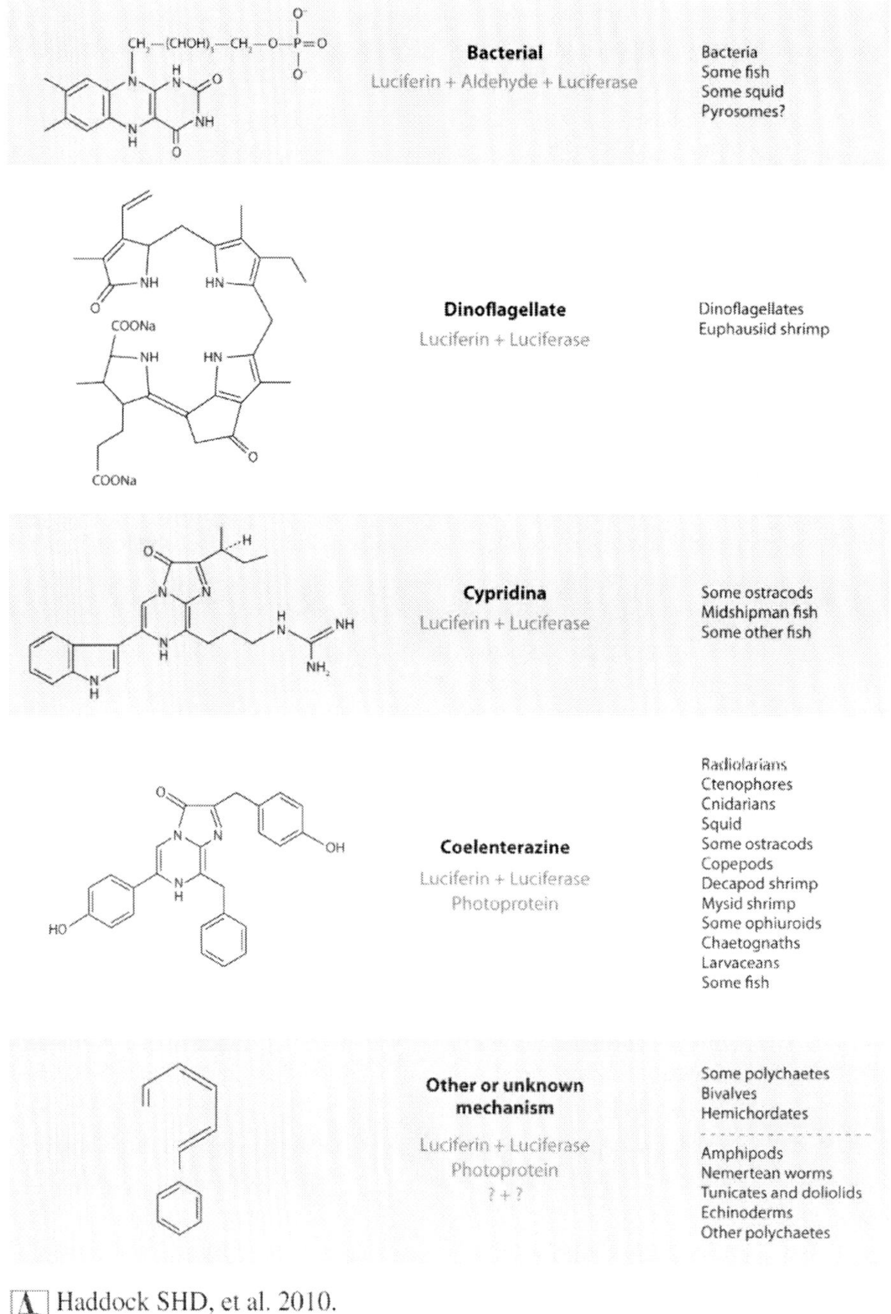

Figure 2. Luciferins used by marine organisms.

Ctenophores use calcium-activated proteins and coelenterazine. Their luminescence can be internally expressed, sometimes in cascading waves as with *Beroe forskalii*, but some species like *Euplokamis stationis*, *Mertensia ovum*, and *Eurhamphaea vexilligera* also emit glowing particles as part of an escape response [Widder *et al*., 1992]. Bioluminescence is very well studied in comb jellies, where more than 90% of planktonic genera have the ability to produce light [Haddock and Case, 1995].

Bioluminescence is found in both benthic and planktonic cnidarians, the group that includes corals, anemones, hydroids, medusae, and siphonophores. All luminous species use coelenterazine as their light-emitting substrate. Luminous hydrozoans include both hydromedusae and siphonophores. Most of the planktonic forms are bioluminescent, including 91% of planktonic siphonophore genera, while for unknown reasons it is rare among certain other groups, like the species of benthic hydroids that do not produce medusae. Most famous of the luminescent hydrozoans, and arguably of all bioluminescent invertebrates, is the shallow-living hydromedusa *Aequorea victoria*, which provided the original source material for research on photoproteins and the Nobel Prize–winning GFP [Shimomura, 2005]. Most hydrozoans likely use bioluminescence for defensive or warning purposes, but siphonophores also use luminescence [Haddock *et al*., 2005] and fluorescence [Pugh and Haddock, 2009] to attract prey directly to their stinging tentacles.

There are many luminous octocorals found in shallow sandy bottoms (*Renilla*, *Ptilosarcus*), and in the deep sea (*Stylatula*, *Halipterus*, *Anthomastus*). Although the hard corals and anemones (Hexacorallia) are now famous for the possession of fluorescent proteins [Shagin *et al*., 2004], they are not usually bioluminescent. Also there are several different bioluminescent lineages in annelids, yet the chemical mechanisms of light production have not been fully known for most species. Some of the terrestrial annelids have been chemically characterized [Petushkov and Rodionova, 2007], but there do not seem to be many parallels between the groups, and luminescence has several independent origins. The life cycles of the famous syllid fireworms, including *Odontosyllis*, have been thoroughly studied through the years. This normally benthic species produces a spawning stage near the time of the full moon. Females produce luminescent secretions that attract the males to swarm around them. Although these polychaetes use bioluminescence during spawning, like most organisms they will also produce internal luminescence in response to physical disturbance [Deheyn and Latz, 2009]. In Eusyllis, fragments can continue luminescing for weeks, even without the head attached [Zörner and Fischer, 2007]. Benthic *Chaetopterus* make light using a unique photoprotein [Shimomura, 2006] five times as large as cnidarian photoproteins. Luminescence is also present in a recently discovered planktonic species of *Chaetopterus* [Osborn and Rouse, 2008].

Bioluminescence makes scattered appearances among the many other wormlike phyla. Several species of acorn worms (Hemichordata) are luminescent, including *Balanoglossus* and the planktonic tornaria larvae of *Ptychodera flava*, whose luciferin was recently characterized [Kanakubo and Isobe, 2005]. Chaetognaths (arrow worms) use a luciferase+coelenterazine chemistry and shed a cloud of glowing particles in conjunction with an escape response [Haddock *et al*., 2010].

Luminous marine molluscs include a few unusual gastropods like the whelk *Planaxis* and the spectacular pelagic nudibranch *Phylliroe*. The most prominent of the bioluminescent marine molluscs are the cephalopods. Among the squids alone, there are at least 70 luminous genera (Herring, 1977). Bacterial symbionts produce luminescence for several genera in the families Sepiolidae and Loliginidae [Nyholm *et al.*, 2009]. The rest of the squids, though, have intrinsic bioluminescence, using a luciferin along with their individual luciferase. Some of these have been chemically characterized. For example, *Symplectoteuthis* has a photoprotein that operates with dehydro-coelenterazine [Isobe *et al.*, 2008]. In *Watasenia scintillans*, the luciferase reacts with coelenterazine-disulfate and also has a requirement for ATP and Mg^{2+} as cofactors, which is unusual for coelenterazine-based luminescence [Tsuji, 2005].

Squids can produce an impressive variety of luminescent displays. The deep-sea vampire squid *Vampyroteuthis* is sufficiently distinct to have been classified in its own order. In addition, two large mantle photophores, and small light organs scattered across its body, it can release glowing particles from special light organs on its arm tips, apparently to distract predators [Robison *et al.*, 2003]. *Octopoteuthis* takes on a variety of coloration patterns, and will drop its arms on being disturbed, leaving the glowing arm tips as decoys [Bush *et al.*, 2009]. Another cephalopod with light organs at its arm tips is *Taningia danae*. This highly active squid has clawlike hooks, and large (up to 2 cm) light organs at its arm tips. They are thought to use luminescence both for intraspecific communication and potentially to stun prey [Kubodera *et al.*, 2007].

Many types of planktonic crustaceans are bioluminescent, and they use species-specific luciferases with at least three different types of luciferins. They have light organs along the lower surface of their body, which they use for counterillumination, and some species also have two small light organs on their eyestalks. These might serve as feedback mechanisms for determining how well their ventral photophores are matching background light. Like most photophores, these are under nervous control, involving serotonin moderated by nitric oxide [Krönström *et al.*, 2007].In Krill, use physical mechanisms are used to contarct and dilate photophores for regulation of light production [Krönström *et al.*, 2009].

Among invertebrates, Copepods, are one of the most abundant bioluminescent groups in the sea, although the widespread genus *Calanus* is not luminescent. Common luminous genera include *Pleuromamma, Metridia, Oncaea*, and *Gaussia*. Bioluminescence involves coelenterazine, exhibited either as intracellular flashes or emitted into the water as part of an escape response [Widder *et al.*, 1999]. GFP-like fluorescent proteins have been cloned from copepods [Masuda *et al.*, 2006], but surprisingly they have only been found in nonluminous species. Many of the bioluminescent copepods are predatory and live moderately deep, although the tiny Poecilostomatoid copepod Oncaea is shallow and pseudoplanktonic [Bottger-Schnack and Schnack, 2005].

Bioluminescence is found in majority of echinoderms: brittle stars (Ophiuroidea), sea stars (Asteroidea), sea cucumbers (Holothuroidea), and even crinoid sea lilies [Herring and Cope, 2005]. Much of the research work on echinoderms has focused on ophiuroid (brittle star) behavior and neurophysiology [Dewael and Mallefet, 2002]. A complex system of neurotransmitters modulates light output in these groups, and light originates from both an unknown photoprotein and a coelenterazine+luciferase reaction, depending on the species [Shimomura, 2006].

In echinoderms other than brittle stars, luminescence is more common among deep-sea taxa (Brisingidae and Paxillosida for sea stars, *Pannychia*, *Peniagone*, and *Scotoanassa* for the holothurians, and *Thaumatocrinus* and *Annacrinus* among crinoids). As is typical for bioluminescence, it is also disproportionately represented in the pelagic holothurians such as *Enypniastes eximia* [Robison, 1992] and *Pelagothuria*. New luminous ophiuroid species continue to be discovered [Mallefet *et al.*, 2004], and our knowledge about diversity of bioluminescent echinoderms will increase with the study of more deep-sea species in good condition.

Atleast 42 families across 11 orders of bony fishes are bioluminescent [Suntsov *et al.*, 2008], in addition to one family of sharks. In contrast with invertebrate taxa, several of these groups including the well-known anglerfishes [Pietsch, 2009], flashlight fish like *Photoblepharon* spp. (Haygood & Distel 1993), and shallow ponyfishes like *Leiognathus* spp. [Wada *et al.*, 1999] use bacterial symbionts for light production. The other luminous fishes use either coelenterazine [Mallefet and Shimomura, 1995], ostracod luciferin, or some other uncharacterized light-emitter, hence having intrinsic luminescence. Fish photophores are often highly modified and adapted to control not only the intensity of light but its angular distribution, depending upon their particular function [Cavallaro *et al.*, 2004].

While a great diversity of marine animals and microbes are able to produce their own light in most of the volume of the ocean, bioluminescence is also the primary source of light. With the exception of some insect larvae, a freshwater limpet, and unsubstantiated reports from deep in Lake Baikal, luminescence is nearly absent in freshwater. On land, fireflies are the most prominent examples, but other luminous taxa exist also including other beetles, insects like flies and springtails, fungi, centipedes and millipedes, a snail, and earthworms. This discrepancy between marine and terrestrial luminescence is not completely understood, but several properties of the ocean support the evolution of luminescence: (*i*) comparatively stable environmental conditions prevail, with a long uninterrupted evolutionary history; (*ii*) the ocean is optically clear in comparison with rivers and lakes; (*iii*) large portions of the habitat receive no more than dim light, or exist in continuous darkness; and (*iv*) interactions occur between a huge diversity of taxa, including predator, parasite, and prey. Given its widespread distribution and signifance, bioluminescence is undoubtly a predominant form of communication in the sea, with signifant effects on the immense daily vertical migration, predator-prey interactions, camouflage, communication, illumination and counterillumination etc. [Haddock *et al.*, 2010].

Bioluminescence is used as a lure to attract prey (Figure-3). This phenomenon was observed most prominently in fish, especially the diverse anglerfishes, which use bacteria to produce a long glow controlled by changng the conditions in the light organ where the bacteria are cultured [Pietsch, 2009]. Many types of stomiid dragonfish also have luminous barbels, although not involving bacteria; in deed, out of 25 genera, there are only two scaleless stomiids without barbels [Kenaley and Hartel, 2005]. Example is *Malacosteus*, which feeds on copepods and has red suborbital photophores, thus suggesting a different strategy for capturing prey.

Attraction through bioluminescence has often been observed for other taxa. In particular, the squid *Chiroteuthis* has specific light organs that dangle at the end of long tentacles hanging down like fishing lines and are thought to serve as lures [Robison *et al.*, 2003]. In one species, the bioluminescent lure is exhibited by a red fluorescent coating [Haddock *et al.*, 2005], enhancing the possibility that this species preys on particular fishes having unexpectedly long-wavelength sensitivity [Turner *et al.*, 2009]. When the fish or squid draws near to attack the prey, it is inturn attacked by the shark. This application of bioluminescence occurs only after the shark is grown, because the ventral pattern of light of another squaloid shark appears to serve a normal counterillumination role in juveniles [Claes and Mallefet, 2008].

Animals may not have to make their own light to gain a predatory advantage from the existence of bioluminescence. Some nonluminous top predators may initiate a burglar alarm response themselves, and actively use bioluminescence in their environment to attract their prey. Ocean sunfish and leatherback turtles both survivemainly on jellies [Hays *et al.*, 2009], and it has been suggested that they use luminescence to find their prey [Houghton *et al.*, 2008].

There are much more examples defensive functions of bioluminescence (Figure-3). For example, when bright flash is evoked at close range, bioluminescence is assumed to upset the predators, in a form of predator intimidation analogous to the peacock butterfly [Vallin *et al.* 2006]. In other case, secretion of a cloud of sparks or glowing fluid makes it difficult for the predator to track the location of its escaping prey. This behavior was observed in many animals, including copepods, shrimp, tube-shoulder searsiid fishes, ctenophores and the vampire squid, which lacks an ink sac but instead emits a cloud of luminous secretions from its arm tips [Robison *et al.* 2003[. Organisms like the deep-sea squid *Octopoteuthis deletron* may autotomize luminous body parts [Bush *et al.*, 2009], which then continue to move and flash to draw away the attention of predators.

Apparently more common, but observed in few cases, is the application of a sacrificial tag. In this situation, an organism may lose part of its body to a predatory encounter. These lost tissues can continue to glow within the predator's stomach for hours afterward [Herring and Widder, 2004; Haddock *et al.*, 2006). This is thought to be the selective force driving the presence of so many black- or red-pigment guts in otherwise transparent animals [Johnsen, 2005], since most red and orange pigments absorb blue light. Many invertebrates also have excellent powers of regeneration and may be able to regrow the missing appendage while their predator suffers from having taken in the "Trojan horse." This phenomenon could be at work in nearly any bioluminescent organism large enough to recover from loss of tissue or skin during an attack.

Bioluminescence is an extremely effective way for invertebrates to communicate to organisms much larger and potentially far away and thus may help explain its prevalence. Depending on the conditions, a bioluminescent flash can be seen from tens to hundreds of meters away [Turner *et al.*, 2009]. Even a single-celled dinoflagellate 0.5 mm long can send a signal to a large fish 5 m away the equivalent of a 2 m tall human being able to communicate over a distance of 20 km. Chemical cues, while certainly playing important and overlooked roles in the sea [Woodson *et al.*, 2007], perform a different set of functions than optical or acoustical signals; they do not diffuse rapidly enough to send the acute signals across distances that are possible through bioluminescence. Chemicals do have the advantage, however, of operating between nonvisual organisms with ineffective bioluminescence. Acoustical signals are transmitted extremely effectively in the sea [Medwin, 2005], but they have the drawback of being relatively nondirectional, and many planktonic organisms lack the hard (or at least firm) body parts or gas bladders required to generate sounds [Henninger and Watson, 2005]. Thus bioluminescence is one of the most effective ways enabling small organism to communicate effectively with much larger organism in the sea.

Communication within species is a well-known function of bioluminescence in terrestrial firefly courtship [Woods Jr *et al.*, 2007], but these types of communication are less well known for the sea. Sexual dimorphism and the use of bioluminescence for mating have been studied by Herring [2007]. Ostracods show species-specific patterns of signaling, complex three-dimensional mate-following behavior [Rivers and Morin, 2008], and even so-called sneaker males that follow along with displaying males and attempt to benefit from the surrounding displays while not producing luminescence of their own [Rivers and Morin, 2009]. Other examples of organisms thought to use bioluminescence for reproduction-related communication are fireworms, the pelagic octopods *Japetella* and *Eledonella*, and the ponyfishes, which produce synchronized group displays [Woodland, 2002] and have evolved luminescent-based sexual dimorphism [Ikejima *et al.*, 2004]. The lures of anglerfish might also be employed for mate-finding purposes in addition to prey attraction [Herring, 2007]. Light emission or visual sensitivity at unusually long or short wavelengths (as is the case with *Japetella* and the firefly squid *Watasenia*) may indicate that bioluminescence is being used as a private communication channel, rather than for one of the interspecific predator-prey functions described above. The presence of species-specific photophore patterns and sexual dimorphism suggests that the organisms use these features to distinguish each other, just as biologists do, but demonstrating this effect is difficult in the deep-sea where many of the organisms are found [Herring, 2007].

On land, it is widely accepted that bright coloration can advertise toxicity or unpalatability, and this has been shown to apply to terrestrial bioluminescence as well [De Cock and Matthysen, 2003]. A similar mechanism has been suggested for many marine organisms, including scale worms, jellyfish, and brittle stars [Herring and Widder, 2004]. It would be particularly suitable for cnidarian bioluminescence to function this way; they are fragile yet potentially deadly, and it is to their advantage and advantageous to other organisms to avoid physical encounters.

Other predatory applications include the use of bright light by a predator to stun or confuse prey. Videos of the squid *Taningia* flashing its tentacles while attacking bait [Kubodera *et al.*, 2007] support this hypothesis, but there is little to no experimental evidence for this. The bright forward-pointing "headlamps" of the myctophid *Diaphus* might operate in this way, although they could also be used to illuminate or to induce fluorescence in prey. The scaleless dragonfishes are thought to use bioluminescence to aid in visual searching, including those with red suborbital photophores [Haddock *et al.*, 2010].

One important characteristic that has been well studied in the laboratory is counterillumination. It is common among crustaceans, cephalopods, and fishes. This form of camouflage involves using ventral (lower) photophores to match the dim light coming from the surface, thus making a potential shadow disappear. Counterillumination can be achieved either through a uniform match to the light field or by sufficiently disrupting the silhouette [Johnsen *et al.*, 2004]. Many predators in the midwater have upwardly directed eyes, sometimes tinted yellow, to search for silhouettes. In *Macropinna microstoma*, the eyes can tilt between upward- and forward-looking positions [Robison and Reisenbichler, 2008], while the spookfish, *Dolichopteryx longipes*, has evolved two fields of view in each eye, one focusing upward with a lens and one looking sideways with a mirror, so it can hunt while keeping an eye on what is happening below [Wagner *et al.*, 2009].

Bioluminescence has revolutionized research into many cellular and molecular-biological processes, ranging from intracellular signalling to gene transcription. It has emerged as an extremely useful and versatile reporter technology. It provides a sensitive, non-destructive, and real-time assay that allows for temporal and spatial measurement. The ability to emit light is dependent on the reducing power of the organism; hence, only metabolically active cells can produce light. The direct relationship between viability and light emission allows the use of bioluminescent bacteria to assess the effect of various chemical, biological, and physical signals [Alloush *et al*., 2006].

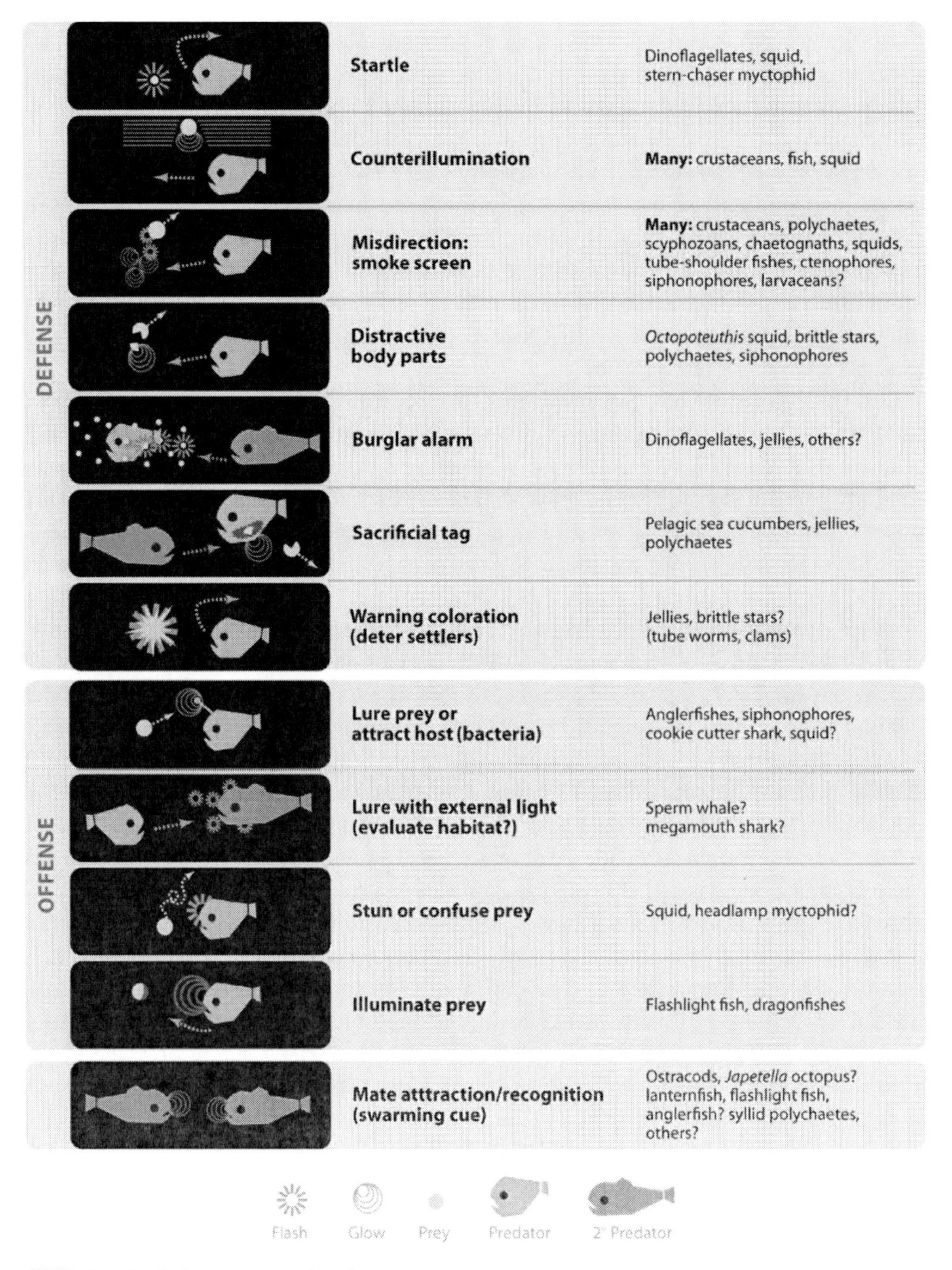

Haddock SHD, et al. 2010.
Annu. Rev. Mar. Sci. 2:443–93

Figure 3. Schematic diagram showing the functions of bioluminescence.

Bioluminescent methods are gaining more and more attention among scientists due to their sensitivity, selectivity and simplicity; coupled with the fact that the bioluminescence can be monitored both in vitro and in vivo. Since the discovery of bioluminescence in the 19th century, enzymes involved in the bioluminescent process have been isolated and cloned. The bioluminescent reactions in several different organisms have also been fully characterized and used as reporters in a wide variety of biochemical assays. From the 1990s it became clear that bioluminescence can be detected and quantified directly from inside a living cell. This gave rise to numerous possibilities for the in vivo monitoring of intracellular processes non-invasively using bioluminescent imaging (BLI) [Brovko and Griffiths, 2007].

Bioluminescence imaging is a powerful methodology that has been developed over the last decade allows a low-cost, noninvasive, and real-time analysis of disease processes at the molecular level in living organisms. It has been used to track tumor cells, bacterial and viral infections, gene expression, and treatment response. *In vivo* bioluminescent imaging (BLI) is a versatile and sensitive tool that is based on detection of light emission from cells or tissues. This form of imaging has successfully applied in mouse models of lung inflammation/injury, bacterial pneumonia, and tumor growth and metastasis [Sadikot and Blackwell, 2005]. Molecular imaging, and specifically bioluminescent reporter gene imaging, has given investigators a high-throughput, inexpensive, and sensitive means for tracking *in vivo* cell proliferation over days, weeks, and even months. The advantage of longitudinal monitoring of both transgene expression and cell survival will provide invaluable information to optimize protocols for clinical cell and gene based therapies. The feasibility to monitor transcriptional regulation and protein–protein interactions will lead to a better understanding of molecular and cellular processes of diseases and might help in accelerating the development and screening of lead compounds. Non-invasive BLI adds spatiotemporal information to our understanding of various disorders and due to the high sensitivity and versatility of this technique the use of BLI will continue to increase. Furthermore, combining anatomical with functional information will broaden even more the applications for *in vivo* monitoring of molecular events in animal models of human disease [Gheysens and Mottaghy, 2009].

Using BLI, researchers were able to identify new features of host–pathogen interactions. Representative examples of BLI applications in bacterial infection are described below. Wiles and colleagues transformed *Citrobacter* with a transposon plasmid to stably integrate the *luxCDABE* operon into the bacterial genome [Wiles *et al.*, 2004]. After confirming that the transformed bacteria were not attenuated *in vivo* or *in vitro* compared with the parent strain, they orally inoculated mice with the bioluminescent *Citrobacter* and monitored colonization and clearance of the mouse colon by BLI and colony counts. These authors made the new discoveries that *Citrobacter* colonized the caecal patch before colonizing the rest of the colon, and bacteria were cleared from the caecum before being cleared from the rest of the colon. Oxygen availability was not a confounding factor in this study because mice were dissected and their intestines moved to a Petri dish prior to imaging. In a subsequent article [Wiles *et al.*, 2005], the same authors use the luminescent *Citrobacter* to show that this bacterium becomes 'hyperinfectious' after passage through a mouse, requiring a lower infectious dose to colonize naïve mice. In addition, these 'hyperinfectious' bacteria did not need to colonize the caecal patch before infecting the rest of the colon.

The development of novel drugs is a lengthy process requiring years of preclinical research and many steps indispensable to ensure that the molecule of interest can be administered to humans with a minimal risk of toxic effects. Even a minimal reduction in the initial stages of drug development would result in a tremendous saving in time; therefore, pharmaceutical companies are eager to apply novel methodologies that shorten the time required for pharmacodynamic, pharmacokinetic and toxicological studies to be carried out *in vitro* and in animal systems. *In vivo* imaging provides surrogate endpoints that can improve the identification of new drug candidates and speed up their research at preclinical stages. The integration of reporter systems in animal models of human diseases represents a reachable frontier that will dramatically advance drug development in terms of costs, time and efficacy [Brovko and Griffiths, 2007].

Bioluminescent organisms can determine toxicity because the noxious substances reduce the glow by killing the bacteria. Bioluminescence technology is a powerful tool for the in situ detection and assessment of sterilization techniques without the need for the lengthy procedures of homogenizing and culturing multiple samples.

Bioluminescent assays can detect a variety of molecules, hormones, proteases, and oligonucleotides. Some of these assays have been incorporated onto miniaturized microfluidic detection platforms. Luciferases in particular are used extensively for mass-market contamination-type assays and in bio-threat detection devices [Roda *et al.*, 2004]. Luciferase detects ATP (which indicates the presence of cells) with such sensitivity that luciferase-based bioluminescence assays have been developed for the analysis of biomass in seawater and water treatment plants; as hygiene monitors in hospitals and food [Moaz *et al.*, 2003] and beverage institutions; for cell viability studies; [Roda *et al.*, 2004] and as reporters for transformation and transfection, gene expression, and promoter assays [Roda *et al.*, 2004]. Further advances are expected with the development of brighter luciferases emitting at different wavelengths and more sensitive detectors. Without doubt, much like our ancestors used bioluminescent foxfire torches and firefly lamps to light their way through dark and unknown territories, researchers will continue discovery of other bioluminescent pigments that emit light at different wavelength to reveal the mystery of the world around us and the depth of life itself.

Acknowledgment

The authors would like to thank 2010 Endeavour research award (Austraining International).

References

Aberg, V; Almqvist, F. Pilicides-small molecules targeting bacterial virulence. *Org Biomol Chem, 2007*, 5, 1827–1834.

Alloush, HM; Lewis, RJ; Salisbury, VC. *Bacterial Bioluminescent Biosensors: Applications in Food and Environmental Monitoring, Analytical Letters, 39*, 1517–1526.

Baker, A; Robbins, I; Moline, MA; Iglesias-Rodriguez, MD. Oligonucleotide primers for the detection of bioluminescent dinoflagellates reveal novel luciferase sequences and information on the molecular evolution of this gene. *J. Phycol,* 2008, 44, 419–428.

Bottger-Schnack, R; Schnack, D. Population structure and fecundity of the microcopepod *Oncaea bispinosa* in the Red Sea - a challenge to general concepts for the scaling of fecundity. *Mar Ecol-Prog Series,* 2005, 302, 159–175.

Brovko, LY; Griffiths, MW. Illuminating cellular physiology: recent developments. *Sci Prog,* 2007, 90, 129-160.

Bush, SL; Robison, BH; Caldwell, RL. Behaving in the dark: locomotor, chromatic, postural, and bioluminescent behaviors of the deep-sea squid *Octopoteuthis deletron* Young 1972. *Biol Bull, 2009,* 216, 7–22.

Campbell, AK; Jean-Marie, Bassot (1933–2007): a life of unquenched curiosity—Obituary. *Luminescence,* 2008, 23, 187–190.

Cavallaro, M; Mammola, CL; Verdiglione, R. Structural and ultrastructural comparison of photophores of two species of deep-sea fishes: *Argyropelecus hemigymnus* and *Maurolicus muelleri. J Fish Biol,* 2004, 64, 1552–1567.

Claes, J; Mallefet, J. Early development of bioluminescence suggests camouflage by counter-illumination in the velvet belly lantern shark *Etmopterus spinax* (Squaloidea: Etmopteridae). *J Fish Biol,* 2008, 73, 1337–1350.

De Cock, R; Matthysen, E. Glow-worm larvae bioluminescence (Coleoptera: Lampyridae) operates as an aposematic signal upon toads (*Bufo bufo*). *Behav Ecol,* 2003, 14, 103–108.

Deheyn DD, Latz MI. 2009. Internal and secreted bioluminescence of the marine polychaete Odontosyllis phosphorea (Syllidae). *Invertebrate Biol.* 128:31–45.

Dewael, Y; Mallefet, J. Luminescence in ophiuroids (Echinodermata) does not share a common nervous control in all species. *J Exp Biol J Exp Biol,* 2002, 205,799–806.

Dunlap, PV; Kita-Tsukamoto, K. Luminous bacteria. *Prokaryotes,* 2006, 2, 863–892.

Gheysens, O; Mottaghy, FM. Method of bioluminescence imaging for molecular imaging of physiological and pathological processes. *Methods,* 2009, 48, 139-145.

Griffith, MW. Applications of Bioluminescence in the Dairy Industry. *J Dairy Sci,* 1993, 76, 3118-3125.

Haddock, SHD; Case, JF. Not all ctenophores are bioluminescent: *Pleurobrachia Biol Bull,* 1995, 189, 356–362.

Haddock, SHD; Dunn, CW; Pugh, PR; Schnitzler, CE. Bioluminescent and red-fluorescent lures in a deep-sea siphonophore. *Science,* 2005, 309, 263.

Haddock, SHD; Moline, MA; Case, JF. Bioluminescence in the Sea. *Annu Rev Mar Sci,* 2010, 2, 443–493.

Haddock, SHD; Rivers, TJ; Robison, BH. Can coelenterates make coelenterazine? Dietary requirement for luciferin in cnidarian bioluminescence. *Proc Natl Acad Sci USA,* 2001, 98, 11148–11151.

Hastings, JW. Biological diversity, chemical mechanisms, and the evolutionary origins of bioluminescent systems. *J Mol Evol,* 1983, 19, 309-321.

Hays, GC; Farquhar, MR; Luschi, P; Teo, SLH; Thys, TM. Vertical niche overlap by two ocean giants with similar diets: Ocean sunfish and leatherback turtles. *J Exp Mar Biol Ecol,* 2009, 370, 134–143.

Henninger, HP; Watson, WH. Mechanisms underlying the production of carapace vibrations and associated waterborne sounds in the American lobster, *Homarus americanus*. *J Exp Biol,* 2005, 208, 3421.

Herring, PJ. Sex with the lights on? A review of bioluminescent sexual dimorphism in the sea. *J Mar Biol Assoc UK,* 2007, 87, 829–842.

Herring, PJ; Cope, C. Red bioluminescence in fishes: on the suborbital photophores of Malacosteus, Pachystomias and Aristostomias. *Mar Biol,* 2005, 148, 383–394.

Herring, PJ; Widder, EA. Bioluminescence of deep-sea coronate medusae (Cnidaria: Scyphozoa). *Mar Biol,* 2004, 146, 39–51.

Houghton, JD; Doyle, TK; Davenport, J; Wilson, RP; Hays, GC. The role of infrequent and extraordinary deep dives in leatherback turtles (*Dermochelys coriacea). J Exp Biol,* 2008, 211, 2566–2575.

Ikejima, K; Ishiguro, N; Wada, M; Kita-Tsukamoto, K; Nishida, M. Molecular phylogeny and possible scenario of ponyfish (Perciformes: Leiognathidae) evolution. *Mol Phylogenet Evol,* 2004, 31, 904–909.

Isobe, M; Kuse, M; Tani, N; Fujii, T; Matsuda, T. Cysteine-390 is the binding site of luminous substance with symplectin, a photoprotein from Okinawan squid, *Symplectoteuthis oualaniensis. P Jpn Acad B Phys,* 2008, 84, 386–392.

Johnsen, S. The red and the black: Bioluminescence and the color of animals in the deep sea. *Integr Comp Biol,* 2005, 45, 234–246.

Johnsen, S; Widder, EA; Mobley, C; 2004. Propagation and perception of bioluminescence: factors affecting counterillumination as a cryptic strategy. *Biol Bull,* 207, 1–16.

Kaeding, AJ; Ast, JC; Pearce, MM; Urbanczyk, H; Kimura, S; Endo, H; Nakamura, M; Dunlap, PV. Phylogenetic diversity and cosymbiosis in the bioluminescent symbioses of "*Photobacterium mandapamensis. Appl Envir Microbiol,* 2007, 73, 3173–3182.

Kanakubo, A; Isobe, M. Isolation of brominated quinones showing chemiluminescence activity from luminous acorn worm, Ptychodera flava. *Bioorg Med Chem,* 2005, 13, 2741–2747.

Kenaley, CP; Hartel, KE. A revision of Atlantic species of Photostomias (Teleostei: Stomiidae: Malacosteinae), with a description of a new species. *Ichthyol. Res.* 2005, 52, 251–263.

Kronstrom, J; Karlsson, W; Johansson, BR; Holmgren, S. Involvement of contractile elements in control of bioluminescence in Northern krill, *Meganyctiphanes norvegica* (M. Sars). *Cell Tissue Res,* 2009, 336, 299–308.

Kronstrom, J; Dupont, S; Mallefet, J; Thorndyke, M; Holmgren, S. Serotonin and nitric oxide interaction in the control of bioluminescence in northern krill, *Meganyctiphanes norvegica* (M. Sars). *J Exp Biol,* 2007, 210, 3179–3187.

Kubodera, T; Koyama, Y; Mori, K. Observations of wild hunting behaviour and bioluminescence of a large deep-sea, eight-armed squid, *Taningia danae. Proc Biol Sci,* 2007, 274, 1029–1034.

Latz, MI; Jeong, HJ. Effect of red tide dinoflagellate diet and cannibalism on the bioluminescence of the heterotrophic dinoflagellates *Protoperidinium* spp. *Mar Ecol Prog Ser*, 1996, 132, 275–285.

Liaqat, I; Bachmann, RT; Sabri, AN; Edyvean, RG. Isolate-specific effects of patulin, penicillic Acid and EDTA on biofilm formation and growth of dental unit water line biofilm isolates. *Curr Microbiol,* 2010, 61, 148-156.

Makemson, JC; Fulayfil, NR; Landry, W; Vanert, LM; Wimpee, CF; Widder, EA; Case, JF. *Shewanella woodyi* sp. nov., an exclusively respiratory luminous bacterium isolated from the Alboran Sea. *Int J Syst Bacteriol,* 1997, 47, 1034–1039.

Mallefet, J; Hendler, G; Herren, CM; McDougall, CM; Case, JF. A new bioluminescent ophiuroid species from the coast of California. In *Echinoderms: M¨ unchen*, ed. T Heinzeller, *JH Nebelsick*, 2004, pp. 305–310.

Mallefet, J; Shimomura, O. Presence of coelenterazine in mesopelagic fishes from the Strait of Messina. *Mar Biol* 1995, 124, 381–385.

Masuda, H; Takenaka, Y; Yamaguchi, A; Nishikawa, S; Mizuno, H. A novel yellowish-green fluorescent protein from the marine copepod, *Chiridius poppei*, and its use as a reporter protein in HeLa cells. *Gene,* 2006, 372, 18–25.

McFall-Ngai, MJ; Ruby, E. Sepiolids and vibrios: when first they meet. *BioScience,* 1998, 48:257–65.

Medwin, H. *Sounds in the sea: from ocean acoustics to acoustical oceanography*. Cambridge, UK: Cambridge Univ Press, 2005, pp. 643.

Moaz, A; Mayr, R; Scherer, S. Temporal stability and biodiversity of two complex antilisterial cheese-ripening microbial consortia. *Appl Environ Microb,* 2003, 69, 4012-4018.

Nealson, KH; Hastings, JW. Quorum sensing on a global scale: massive numbers of bioluminescentbacteria make milky seas. *Appl Environ Microbiol,* 2006, 72, 2295–2297.

Nyholm, SV; McFall-Ngai, M. 2004. The winnowing: establishing the squid-vibrio symbiosis. *Nat Rev Microbiol,* 2, 632.

Nyholm, SV; Stewart, JJ; Ruby, EG; McFall-Ngai, MJ. Recognition between symbiotic *Vibrio fischeri* and the haemocytes of *Euprymna scolopes*. *Environ Microbiol,* 2009, 11, 483–493.

Osborn, KJ; Rouse, GW. Multiple origins of pelagicism within Flabelligeridae (Annelida). *Mol Phylogenet Evol,* 2008, 49, 386–392

Petushkov, VN; Rodionova, NS. Purification and partial spectral characterization of a novel luciferin from the luminous enchytraeid *Fridericia heliota*. *J Photochem. Photobiol,* 2007, 87, 130–136.

Pietsch, TW. *Oceanic Anglerfishes: Extraordinary Diversity in the Deep Sea*. Berkeley: Univ of Calif Press, 2009, 576 pp.

Pugh, PR; Haddock, SHD. Three new species of Resomiid siphonophores (Siphonophora, Physonectae). *J. Mar. Biol. Assoc.* UK, 2009.doi: 10.1017/S0025315409990543

Rivers, TJ; Morin, JG. Plasticity of male mating behaviour in a marine bioluminescent ostracod in both time and space. *Animal Behav,* 2009, 78, 723-734.

Robison, BH. Bioluminescence in the benthopelagic holothurian *Enypniastes eximia*. *J Mar Biol Assoc UK,* 1992, 72, 463–472.

Robison, BH; Reisenbichler, KR. *Macropinna microstoma* and the paradox of its tubular eyes. *Copeia,* 2008, 4, 780–784.

Robison, BH; Reisenbichler, KR; Hunt, JC; Haddock, SHD. Light production by the arm tips of the deep-sea cephalopod *Vampyroteuthis infernalis*. *Biol Bull;* 2003; 205; 102–109.

Roda, A; Pasini, P; Mirasoli, M; Michelini, E; Guardigli, M. Biotechnological applications of bioluminescence and chemiluminescence. *Trends Biotechnol,* 2004, 22, 295-303.

Ruby, E; Urbanowski, M; Campbell, J; Dunn, A; Faini, M; Gunsalus, R; Lostroh, P; Lupp, C; McCann, J; Millikan, D; Schaefer, A; Stabb, E; Stevens, A; Visick, K; Whistler, C; Greenberg, EP. Complete genome sequence of *Vibrio fischeri*: a symbiotic bacterium with pathogenic congeners. *Proc Natl Acad Sci USA,* 2005, 102, 3004–3009.

Sadikot, RT; Blackwell, TS. Bioluminescence imaging. *Proc Am Thorac Soc,* 2005, 2, 537-540.

Shagin, DA; Barsova, EV; Yanushevich, YG; Fradkov, AF; Lukyanov, KA; Labas, YA; Semenova, TN; Ugalde, JA; Meyers, A; Nunez, JM; Widder, EA; Lukyanov, SA; Matz, MV. GFP-like proteins as ubiquitous metazoan superfamily: evolution of functional features and structural complexity. *Mol Biol Evol,* 2004, 21, 841–850.

Shimomura, O. *Bioluminescence: Chemical Principles And Methods.* Singapore:World Scientific, 2006, 500 pp.

Shimomura, O. The discovery of aequorin and green fluorescent protein. *J Microsc,* 2005, 217, 3–15.

Suntsov, AV; Widder, EA; Sutton, TT. Bioluminescence in larval fishes. In *Fish Larval Physiology,* ed. RN Finn, BG Kapoor, pp. 2008, 51–88. Bergen, Norway: Univ. Bergen Press.

Tsuji, FI. Role of molecular oxygen in the bioluminescence of the firefly squid, *Watasenia scintillans*. *Biochem Biophys Res Commun*, 2005, 338, 250–253.

Turner, JR; White, EM; Collins, MA; Partridge, JC; Douglas, RH. Vision in lanternfish (Myctophidae): adaptations for viewing bioluminescence in the deep-sea. *Deep Sea Res,* 2009, I56, 1003–1017.

Vallin, A; Jakobsson, S; Lind, J; Wiklund, C. Crypsis versus intimidation—anti-predation defence in three closely related butterflies. *Behav Ecol Sociobiol,* 2006, 59, 455–459.

Wagner, H; Douglas, RH; Frank, TM; Roberts, NW; Partridge, JC. A novel vertebrate eye using both refractive and reflective optics. *Curr Biol*, 2009, 19, 108–114.

Widder, EA; Greene, CH; Youngbluth, MJ. Bioluminescence of sound-scattering layers in the Gulf of Maine. *J Plankton Res,* 1992, 14, 1607–1624.

Widder, EA; Johnsen, S; Bernstein, SA; Case, JF; Neilson, DJ. Thin layers of bioluminescent copepods found at density discontinuities in the water column. *Mar Biol,* 1999, 134, 429–437.

Wiles, S; Clare, S; Harker, J; Huett, A; Young, D; Dougan, G; and Frankel, G. Organ specificity, colonization and clearance dynamics *in vivo* following oral challenges with the murine pathogen citrobacter rodentium. *Cell Microbiol,* 2004, 6, 963–972.

Wiles, S; Dougan, G; Frankel, G. Emergence of a 'hyperinfectious' bacterial state after passage of citrobacter rodentium through the host gastrointestinal tract. *Cell Microbiol,* 2005, 7, 1163–1172.

Wilson, T; Hastings, JW. Bioluminescence. Annu Rev Cell Dev Biol, 1998, 14, 197-230.

Woodland, DJ; Cabanban, AS; Taylor, VM; Taylor, RJ. A synchronized rhythmic flashing light display by schooling *Leiognathus splendens* (Leiognathidae: Perciformes). *Mar Freshwater Res,* 2002, 53, 159–162.

Woods, WJr; Hendrickson, H; Mason, J; Lewis, S. Energy and predation costs of firefly courtship signals. *Am Nat,* 2007, 170, 702–708.

Woodson, CB; Webster, DR; Weissburg, MJ; Yen, J. Cue hierarchy and foraging in calanoid copepods: Ecological implications of oceanographic structure. *Mar Ecol Prog Ser,* 2007, 330, 163–177.

Zorner, SA; Fischer, A. The spatial pattern of bioluminescent flashes in the polychaete *Eusyllis blomstrandi* (Annelida). *Helgoland Mar Res*, 2007, 61, 55–66.

In: Bioluminescence
Editor: David J. Rodgerson, pp. 115-127
ISBN 978-1-61209-747-3

Chapter 6

BIOLUMINESCENCE RESONANT ENERGY TRANSFER AS A BASIS FOR A NOVEL HYPOXIA BIOSENSOR FOR *IN VIVO* FUNCTIONAL IMAGING

P. Iglesias and Jose A. Costoya
Molecular Oncology Laboratory MOL,
Departamento de Fisioloxia, Universidade de Santiago de Compostela,
Santiago de Compostela, Spain

ABSTRACT

Optical imaging methods such as fluorescence are provided with a broad range of proteins and dyes used to visualize many types of these biological processes widely used in cell biology studies and lately also in clinical research. Although the best-known example of these fluorochromes is the green fluorescent protein (GFP), tissue autofluorescence and signal dispersion raise doubts about its suitability as an *in vivo* tracer. Bioluminescence, on the other hand, does not show this signal attenuation caused by living tissues but relies on a chemical reaction for bioluminescent light to be emitted. Taking into account all of these characteristics of the two most known and used optical methods we have devised a novel biosensor comprised by a dual fluorescence-bioluminescence tracer activatable only when the intracellular oxygen concentrations are low enough, or hypoxia. This fusion protein is able to display both fluorescent and bioluminescent properties besides of auto-excitation through a bioluminescent resonance energy transfer or BRET phenomenon, which allows for a better and easier fluorescence performance in localizations where some tissues tend to hinder the fluorochrome excitation.

Keywords: bioluminescence, fluorescence, BRET, hypoxia, biosensor

INTRODUCTION

Imaging Techniques in Oncology

Imaging has become a critical tool in oncology, especially in the visualization of tumoral processes, either for disease detection and diagnostic purposes or as part of clinical trials and basic research. In oncology, the different imaging techniques available can be divided in three different areas depending on its properties. The first class, anatomical imaging, would be comprised by techniques used to locate and measure the size of the tumor, and also facilitating the assessment of the progression of the disease and the effectivity of the therapy. The most popular anatomical techniques are magnetic resonance (MRI) and X-ray computed tomography (CT), which have been playing a prominent role in the last decades since their debut as diagnostic tools. However, these techniques fail to give a broader insight into the physiological features of the tumoral mass and the quality of these images is seriously affected if the density of the tumor and its surroundings is similar [1]. A second class would be the functional imaging techniques, which give more detailed information about tumoral physiological processes such as oxygenation rate, perfusion and alterations of blood flow. To this regard, a variation of the MRI technique, functional MRI (fMRI) is currently being employed as a noninvasive method to preoperatively map functional cerebral cortex and to identify eloquent areas of the cerebral cortex in relation to brain cancers [2]. Finally, molecular imaging techniques have been the last to join but they are rapidly gaining momentum in popularity. Formally, they can be defined as "the visualization, characterization, and measurement of biological processes at the molecular and cellular levels in humans and other living systems" [3], i.e. these techniques inform on the molecular mechanisms underlying the biological processes of interest, which in our case are those that lead to tumoral progression[2, 4, 5]. Recently, multimodal imaging has been on the rise since by combining several techniques a much more complete information about the spatial localization and tumor pathophysiology can be obtained with a dramatic increase in specifity and sensitivity. Examples of these multimodal techniques are Positronemissiontomography (PET)–MRI, PET–CT and MRI–optical (fluorescence) [6-10].

Optical Methods Overview

Some of these techniques rely on ionizing radiation, such as CT and SPECT, involving higher doses than common X-ray imaging procedures, increasingexposuretoradiation in thepopulationwhichsometime in thenearfuturemight be considered a publichealthissue[11]. Besides, while the resolution of MRI and its ability to confer anatomic detail are difficult to match, this technique also requires extremely expensive instrumentation and is time-consuming, giving a poor throughput performance. On the other hand, optical methods such as fluorescence and bioluminescence rely on the emission of visible light that, unlike the former ones do not display the harming effects of ionizing radiations on living organisms or require high-budget equipments to monitor the overall process [12].

In the last years, fluorescence has been rising to a position of prominence in molecular biology thanks to the widespread use of the *Aequoerea victoria* green fluorescent protein

(GFP) [13,14] that as of today remains as the most common molecular reporter. Since then, it has been developed a whole array of new fluorescent proteins with diverse excitation and emission wavelengths that comprises almost the whole visible spectrum, making this technique very adaptable for a wide range of applications [15]. In addition, these novel fluorescent proteins display interesting features such as NIR-shifted emission wavelengths that permit avoiding overlapping parasite emissions from tissue and/or organic compounds [16,17].

In the same manner, luciferase enzymes are commonly employed as reporter genes in cell and molecular biology [18]. Bioluminescence is produced by luciferases that catalyze 'light-emitting reactions' by oxygenating a substrate molecule, luciferin in our case. This process occurs in a number of living organisms, vertebrates, invertebrates and microorganisms. Unlike fluorescence where electron excitation and subsequent photon emission is mediated by light absorption, bioluminescence chemically produces singlet state species that subsequently decay emitting photons of visible light. Sources of luciferases are insects such as the firefly *Photynuspyralis*, marine invertebrates (*Renillareniformis*), plants (*Gaussiaprinceps*) and microorganisms such as several species of vibrionaceae [19]. Although fluorescent light is usually brighter, with less light scattering and photon attenuation that makes fluorescence more suitable for 3D reconstruction, bioluminescent light lacks the problem of cell and tissue auto-luminescence, fluorescent photobleaching and in general facilitates quantitative imaging in deeper localizations than fluorescence. In addition, either luciferase or its substrate D-luciferin seem to be innocuous to living organisms and at the same time this substrate rapidly becomes available and even crossing the blood-brain barrier immediately upon intraperitoneal or intravenous administration [20].

Bioluminescence Resonant Energy Transfer (BRET)

Transference of resonant energy is a well-known phenomenon on which rely proteomic and biochemical procedures such as determination of protein-protein interactions [21]. In the case of BRET, this transference takes place between a luminescent donor (luciferase) and a fluorochrome that acts as acceptor. The energy transfer is strictly dependent on the proximity of both donor and acceptor, being the optimal distance in the range of 1-10 nm [22]. Although the choice of the suitable donor/acceptor pair is usually determined empirically, one of the most popular pairs is *Renilla* luciferase/GFP since this pair exhibits an appropriate spectral overlap of donor emission and acceptor excitation, which is one of the critical steps on the overall performance of the system. As mentioned above, other important aspect to consider is the distance between donor and acceptor, and in the particular case of fusion proteins between both proteins where obviously both proteins are close enough for the energy transference to occur, it is crucial to attend to the freedom of movement necessary between the two moieties of the fusion protein to allow a suitable spatial orientation for BRET to occur. This orientation and freedom of movement can also be empirically tuned by inserting flexible linkers in-between both [23].

This physical phenomenon may be used as a basis of a biosensor as the self-illuminating quantum dots conjugated with luciferase [24]. Quantum dots are semiconductor nanocrystals with different optical properties depending on its size and its composition [25]. In this case, the authors conjugated several copies of a variant of *R. reniformis* luciferase (emission = 480 nm) to several polymer-coated ZnS/CdSe core shell quantum dots that emit fluorescent light at 655, 705 and 800 nm. Also, and in order to increase the cell uptake of these nanostructures, these functionalized quantum dots were conjugated with a polycationic peptide. The system proved to be highly effective in all cases (655, 705, 800 nm) producing a quantifiable BRET signal thus demonstrating the possibilities of BRET-based biosensors modulated by specific biological interactions.

Optical Methods in Clinical Surgery

These techniques are expected to fully develop in the forthcoming years as a prognostic tool for the clinical environment and more specifically for cancer treatment. In this context, to ensure a full recovery and to minimize the risk of recurrence is required the complete removal of the tumor prior to metastasis occurrence. To this regard, and in order to detect the malignant cells conforming the tumoral mass several cancer hallmarks have been used to design targeted molecular probes: increased growth (augmented production of growth factor and growth factor receptors), unrestricted replicative potential, sustained angiogenesis and invasiveness of neighboring tissues and/or metastatic abilities [26]. However, a different approach can be followed by using activatable probes that focus on proteases found in a relative abundance in the surroundings of the tumor and are associated with one of the tumoral features mentioned above (e.g. metalloproteinases as metastatic potential markers). These probes are administered in a quenched state, only displaying a basal emission until cleaved by its target protease. Examples of these activatable probes are those developed by Weissleder and colleagues [27,28] and currently commercialized by VisEn Medical (ProSense and AngioSense).

The first steps for translation of this technique to the clinic have been made in sentinel lymph node mapping. The presence of cancer cells in regional lymph nodes indicates metastasis and necessitates more aggressive, systemic treatment, such as chemotherapy. In many cancer surgeries, oncologists remove several lymph nodes in the region of the tumor to detect metastasis through the lymphatic system. As of today, several initiatives have been made to translate optical methods to clinical imaging as an intraoperative aid for surgeons, even though they are still in early phases of development. One of these initiatives involves intraoperative near-infrared fluorescence monitoring employing low weight molecular ligands (peptides and small molecules) able to target tumoral cells in their niches. This approach is currently being assessed as prospective intraoperatory assistance to surgeons [29,30] with NIR-emitting derivatives of indocyanine green (ICG).

Hypoxia as a TumoralAggresivity Marker in Cancer

One of the most recognizable features of a tumoral cell is the chaotic growth that is intimately related to tumor aggressiveness and invasiveness. As normal cells do, these tumoral cells secrete angiogenic signals to attract additional blood supplies as a response to hypoxia, which usually ensues as soon as the tumor enlarges beyond a millimeter or two in diameter. As a consequence of the low levels of oxygen (hypoxia), the angiogenic switch of these tumoral cells is activated resulting in secretion of hypoxic transcription factors and the number of blood vessels supporting the tumor rises exponentially to fulfill the exacerbated need of nutrients and oxygen [31].

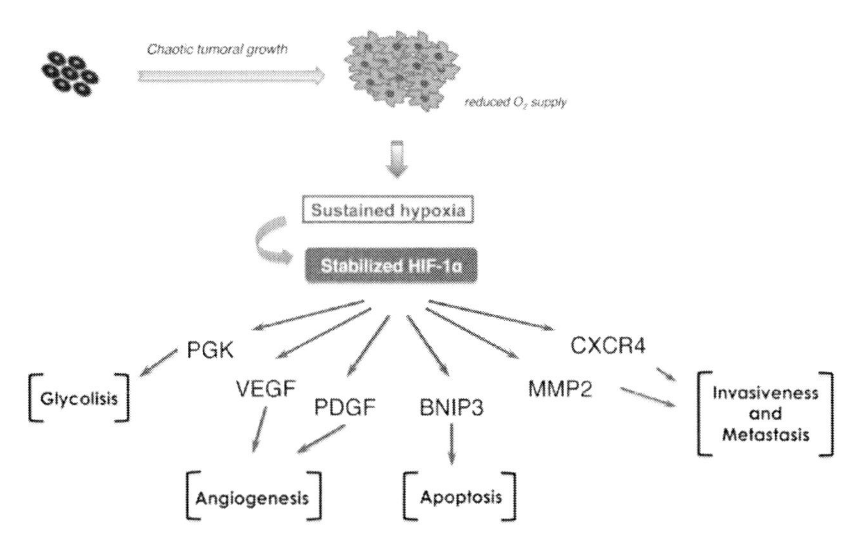

Figure 1. Different outcomes of the biological functions of the transcription factor HIF-1α. This transcription factor is involved in processes such as glycolysis, cell growth and death, and angiogenesis.

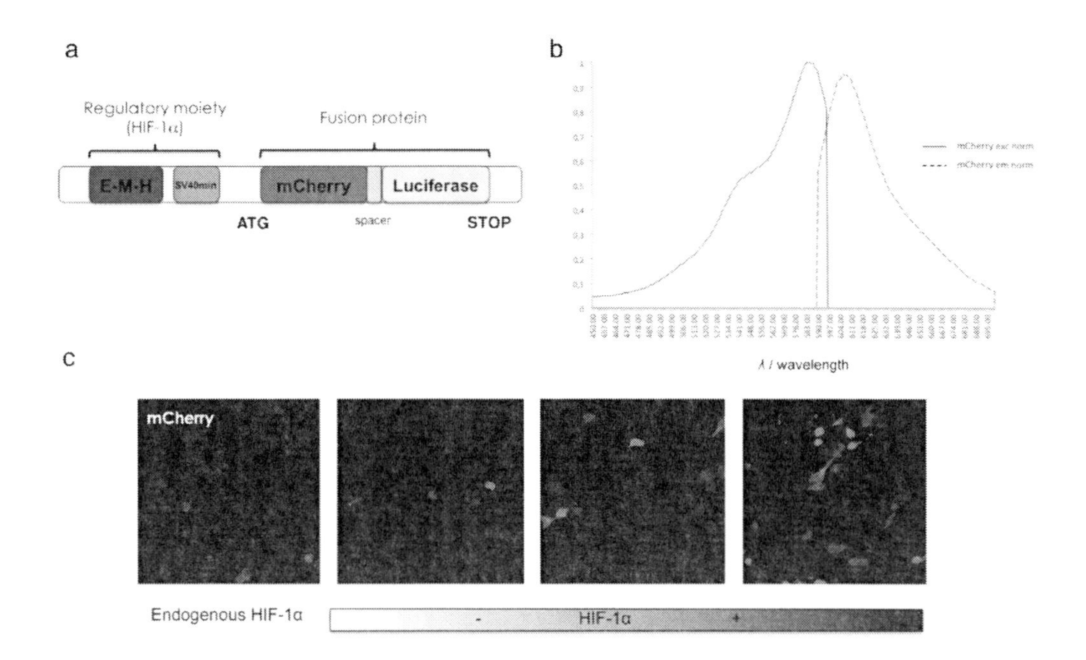

Figure 2. a, Scheme showing the design of the genetically encoded biosensor: a regulatory module formed by the E-M-H enhancer and the SV40 minimal promoter, and the tracer module comprised by the fusion protein formed by the mCherryfluorophore and the firefly luciferase; b, spectrophotometric profile of mCherry in the fusion protein. Solid line represents mCherry excitation while dashed line represents mCherry emission; c, fluorescence images of the mCherry protein.

Attending to hystopathological analyses it is not uncommon to find necrotic regions inside the tumoral core, which is densely populated, in highly proliferative tumoral phenotypes. In order to alleviate this deficiency, tumoral cells elicit the formation of neovessels by stimulating neoangiogenesis. This angiogenic process is tightly regulated and results in the participation of several transcription factors, being HIF-1α one of the most important. This factor appears stabilized in tumors with a high demand of nutrients and oxygen [32]. As Figure 1 shows, this transcription factor exerts its transcriptional activity on several target genes that intervene in crucial processes for the tumoral phenotype such as glycolysis, apoptosis and metastasis [33],making it a good marker of tumoral aggressiveness and invasion capacity in several types of tumors [34-37].

The HIF-1αspecifity also poses another advantage when compared to other constructs harboring entire promoter regions of target genes regulated by this transcription factor such as VEGF [38]. Numerous physiological and pathological processes regulate VEGF expression, such as some cytokines and growth factors signaling pathways that are known to modulate VEGF expression at transcriptional level: IL-lβ, IL-6, PDGF-BB, TGF-β, basic fibroblast growth factor, EGF and HGF are some good examples of the promiscuity of this promoter [39].

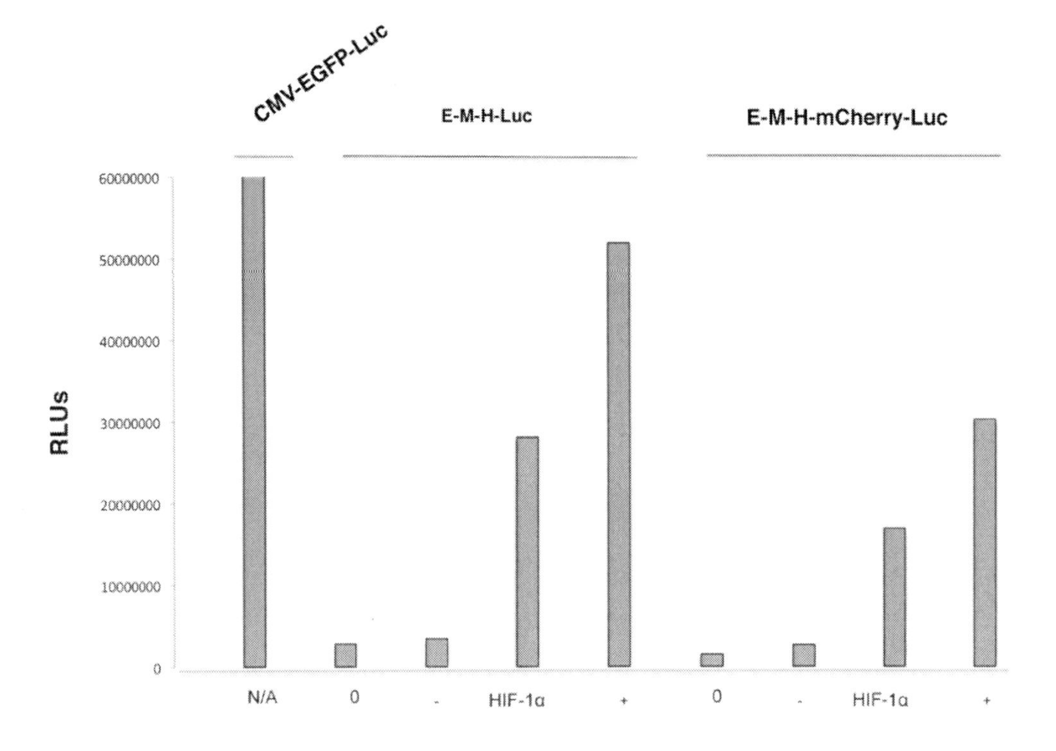

Figure 3. Luciferase activity of the fusion protein for E-M-H-Luciferase and E-M-H-mCherry-Luciferase.

RESULTS

In this review, we describe the design, construction and characterization of a novel hypoxia genetic biosensor with near-infrared fluorescence (NIRF) and bioluminescent properties. This genetic biosensor comprises a regulatory moiety activated by the hypoxia inducible factor HIF-1α, enabling the transcription of a fusion protein that acts as a dual fluorescence-bioluminescence tracer capable of BRET-mediated fluorochrome excitation. All of these data and the corresponding materials and methods employed have been previously described [40].

The structure of this genetic biosensor is outlined in Figure 2a. The regulatory moiety is formed by a novel chimeric enhancer able to recruit the alpha subunit of the HIF-1 transcription factor in a more efficient fashion than the canonical hypoxia response elements (HRE). This chimeric enhancer comprises the (Egr-1)-binding site (EBS) from the Egr-1 gene, the metal-response element (MRE) from the metallothionein gene, and the hypoxia-response element (HRE) from the phosphoglycerate kinase 1 gene. This enhancer has been described to trigger a transcriptional response to a greater extent and increases hypoxia responsiveness than to that of the hypoxia-response element (HRE) alone [41]. A SV40 minimal promoter is located downstream of the chimeric enhacer E-M-H.

On the other hand, the tracer moiety is a dual tracer formed by a fusion protein of a NIR fluorophore (mCherry) and the firefly (*P. pyralis*) luciferase. Several fluorescent proteins were considered on the basis of their excitation and emission wavelengths. Although initially mPlum were tested as prospective fluorochrome, it was eventually discarded due to its low brightness (data not shown and [42]). On the other hand, mCherry presents an excitation wavelength of 585 nm, which makes this fluorochrome an ideal acceptor of bioluminescent light (575 nm), as well as its near NIR-emission avoids autofluorescence phenomena occurring in living tissue.

Since the integrity of the aminoacidic environment of the fluorescent proteins is crucial in order to maintain the optical [43-45], we wanted to assess the fluorescent and bioluminescent activity of the fusion protein in order to disregard any sequence discrepancy with the previously reported excitation/emission wavelengths. Accordingly, we registered the spectrophotometric profiles of mCherry and the firefly luciferase. Figure 2b shows that mCherry displays the same excitation/emission wavelengths that that of the ones reported before [15], indicating that fusing the luciferase and mCherry together did not affect the *in vitro* performance of the fluorescent protein.

Upon determining that the fusion protein retained its optical properties, we wanted to corroborate these data by testing the *in vivo* functionality of the cloned fluorescent protein (Figure 2c). We next transfected our vector (E-M-H-mCherry-Luc) along with increasing amounts of the HIF-1α transcription factor (encoded by pcDNA3-HIF-1α) into the HEK 293 cell line. As expected, those transfected cells displayed a proportional fluorescent signal according to the amounts of transcription factor transfected present in each assay. Thus, our data indicate that the system is proportionally responsive to the amount of transcription factor transfected in each case. Curiously, although the basal induction (i.e. induction by endogenous levels of HIF-1α) proved to be higher than the lowest amount of HIF-1α, the luciferase assay indicates that the response elicited in both cases is similar, thus hinting that the threshold of detection corresponds to the amount of HIF-1α transfected in this case.

In the same way, we next tested the luciferase activity and performance of the system *in* vitro. As shown in Figure 3 and in keeping with the previous results, bioluminescent light appears to be proportionally higher to the concentration of the transcription factor HIF-1α. As a control, it was included a group transfected with the empty vector that lacks the fluorescent protein, and intriguingly it shows a similar response but with a significantly higher maximum value than to that of the biosensor, hinting to a probable energy transfer between the luciferase and mCherry. As means of transfection normalization and to rule out discrepancies on transfection efficiency in the different experimental groups, all data were normalized against the β-galactosidase activity of each group.

So far, these data show that the transcription factor HIF-1α binds efficiently to the response elements located in the chimeric enhancer of the biosensor eliciting the transcription of the tracer fusion protein mCherry-luciferase. Also, although the fusion protein retains its expected fluorescent properties and the luciferase activity it also seems to be lower than to that of the vector containing only the response element and the luciferase (E-M-H-Luc). This difference will be further investigated to determine whether it is caused by a phenomenon of transference of resonant energy between the luciferase and the fluorescent protein, or by a non-expected hindrance of the luciferase catalytic site caused by mCherry resulting in a lower catalytic activity.

Subsequently, we next investigated whether or not the *in vitro* performance of the system could be replicated in an *in vivo* environment. To test this hypothesis, we co-transfected HEK 293 cells again with our biosensor and either pcDNA3-HIF-1α, or a transfection control. We obtained two groups of cells that contained either the activated system (biosensor + HIF-1α) or 'basal state' non-activated cells. Both groups were injected subcutaneously in the hindquarters of immunodeficient SCID.

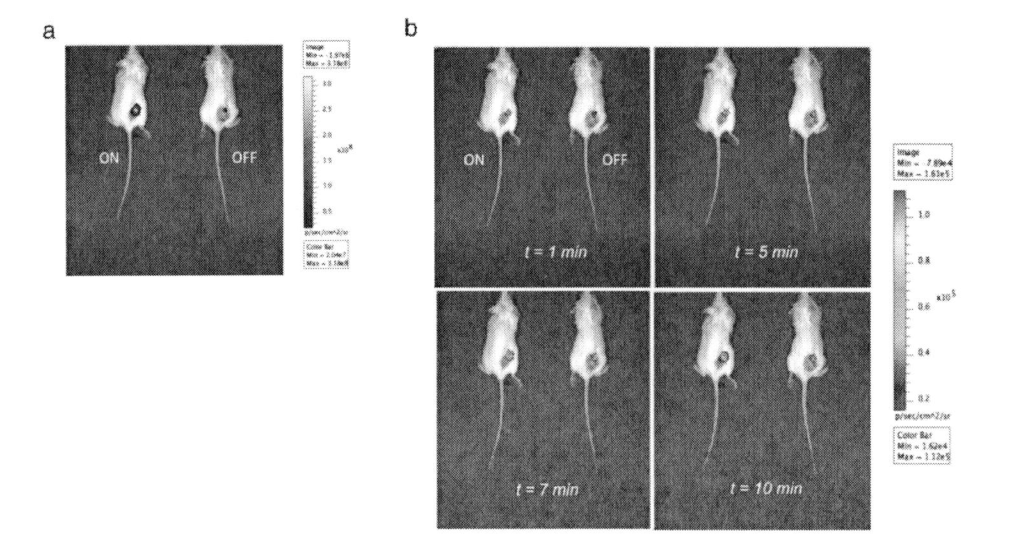

Figure 4. a, *in vivo* fluorescence measure in a xenograft implanted subcutaneously in SCID mice: mCherry fluorescence in the 'activated system' (ON) in HEK 293 cells transfected with E-M-H-mCherry-Luciferase and pcDNA3-HIF-1α and 'non-activated' system (OFF) in cells transfected with E-M-H-mCherry-Luciferase and pcDNA3; b, *in vivo* luciferase activity in SCID mice at various times of acquisition (1, 5, 7 and 11 minutes upon injection D-Luciferin).

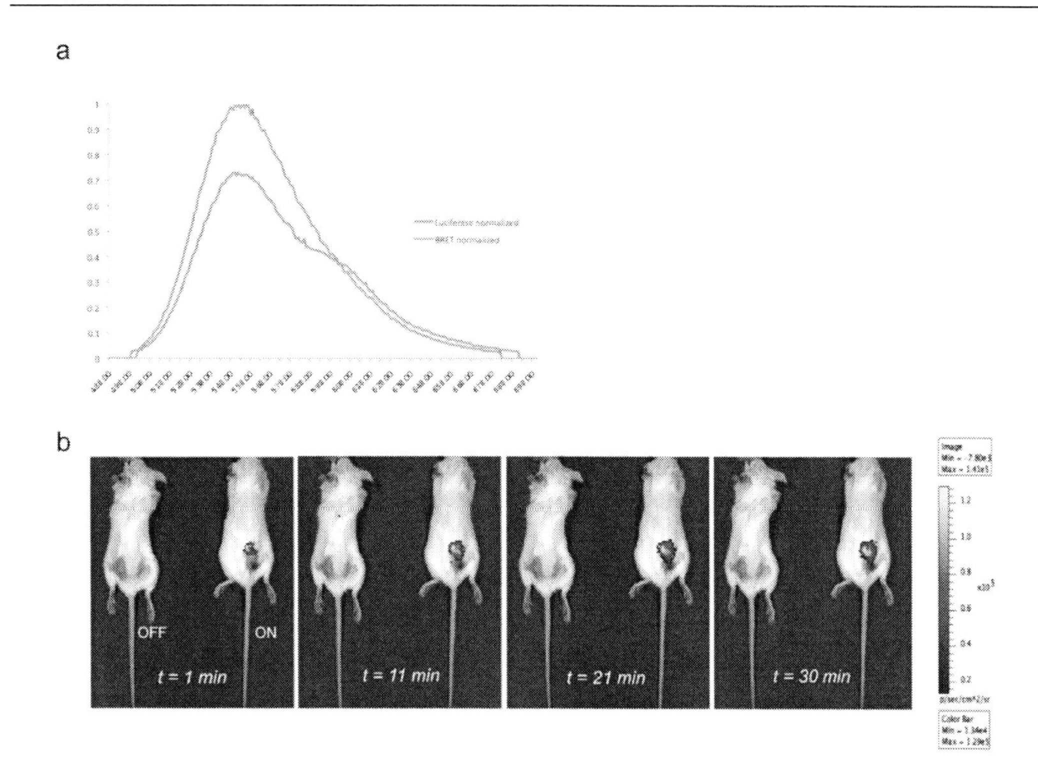

Figure 5. a, *In vitro* BRET performance of the genetically encoded biosensor. Solid line represents luciferase activity of the E-M-H-Luciferase vector while dashed line represents luciferase activity of the E-M-H-mCherry-Luciferase vector; b, *in vivo* BRET performance of the system at various times of acquisition (1, 11, 21 and 30 minutes uopn injection of D-Luciferin).

Figure 4a shows the fluorescent signal measured *in vivo* 24 hours upon injection of both cell populations. In keeping with previous data, we corroborated that the system is active in HEK 293 cells transfected with our biosensor upon subcutaneous injection, and that its intensity is directly proportional to the quantity of the transcription factor HIF-1α present in those cells, as shown by the signal differences between the 'activated' and the 'non-activated' cells. Likewise, luciferase activity was also registered *in vivo* (Figure 4b). Upon administration of luciferin, we observed a similar situation as with fluorescence where both cell populations displayed bioluminescent signals but again a higher intensity in the case of cells with the activated system.

Taken together, these data demonstrate that our hypoxia biosensor is able to proportionally induce the transcription of the mCherry-luciferase tracer when the concentration of HIF-1α is high enough to bind the response elements located upstream the fusion protein coding sequence. Moreover, a similar response was observed both *in vitro*, in HEK 293 cells transfected with increasing concentrations of HIF-1α, and *in vivo*xenografts of these transfected cells, as shown in Figure 4a and Figure 4b.

As discussed above, it was observed a fall in the luciferase activity of the system when compared to the vector that only contains the response element and the luciferase, E-M-H-Luc (Figure 3). We wanted to prove whether this difference can be attributed to a BRET-mediated mCherry excitation in the absence of an external source, or as we mentioned before it is the result of a hindered and defective luciferase moiety. Consequently, we first tested whether or not this transference was taking place *in vitro* by comparing the spectrophotometric profiles of whole cell lysates of cells transfected with the parental vector (E-M-H-Luc) or with our biosensor (E-M-H-mCherry-Luc). As Figure 5a shows, the luciferase alone displays a maximum value at the expected wavelength of 575 nm. However, it also displays a lower second peak at a wavelength corresponding to that of mCherry emission maximum wavelength. In addition, the decrease of the bioluminescent activity was also observed in this case and that would be consistent with the existence of BRET as suggested by the second emission peak as part of the bioluminescent light would be absorbed by the fluorochrome rather than registered by the spectrofluorometer detector.

We finally investigated whether or not this BRET phenomenon could be also detected *in vivo*. For that purpose, we registered the fluorescent emission of the activated system while blocking the excitation filter in order to avoid the interference of any external excitation source. Since this process critically depends on the light emitted by the reaction catalyzed by the luciferase only the fluorescent emission registered originated by BRET was registered upon the injection of the luciferase substrate. Figure 5b shows the most representative points of the series including the peak emission of BRET in SCID mice, reached at 30 minutes upon luciferin injection.

Conclusion

Functional imaging is one of the multiple modalities of imaging in oncology. Optical methods such as fluorescence and bioluminescence can be grouped under this classification when they target a characteristic biological process that in our case is tumoral progression. As an example, we have described the designing and development of a hypoxia-sensing system with both NIR fluorescence and bioluminescence properties. This genetically encoded biosensor is induced by the hypoxia transcription factor HIF-1α, which acts as key regulator of the angiogenic switch of tumoral cells in response to hypoxic conditions. We have demonstrated that this factor binds to the response element located in the regulatory module of the construction efficiently inducing the transcription of the fusion protein, displaying at the same time a proportional response to the concentration of HIF-1α within the cells carrying the biosensor. This HIF-1α specificity also poses another advantage when compared to other constructs harboring entire promoter regions of target genes regulated by this transcription factor such as VEGF, whose outcomes affect multiple processes [39]. We also characterized its fluorescent and bioluminescent properties both in vitro and *in vivo*, observing a BRET phenomenon occurring between the firefly luciferase (donor) and the mCherry protein (acceptor). Although the development of an optical biosensor [46,47] or even a BRET-based biosensor [24] is not a novelty in the field, combining NIR fluorescence and bioluminescence results in a valuable alternative approach for future inducible biosensors that take advantage of BRET. This is especially true in scenarios where an internal excitation source such as intracranial locations where fluorescence alone is not able to overcome the signal attenuation [48].

ACKNOWLEDGMENTS

We thank the members of Molecular Oncology Laboratory MOL for helpful discussions. We also thank M.E. Vazquez for helpful discussions and assistance with the spectrophotometric analysis. We also thank Prof. R. Y. Tsien and Dr. W. H. Suh for kindly providing us with some of the thereagents used in our study. This study was supported by the Spanish Ministry of Education and Science SAF2008-00543 and SAF2009-08629, Xunta de Galicia INCITE08PXIB208091PR (JAC) and by Fundacion de Investigacion Medica Mutua Madrileña (J.A.C., P.I.). Some of the figures were adapted fromBiosensors & Biolectronics, 10, Iglesias, P. And Costoya, J.A., A novel BRET-based genetically encoded biosensor for functional imaging of hypoxia, 13126-13130, Licensenumber 2551330657045 (2010), with permission from Elsevier.

REFERENCES

[1] Seaman ME et al. Molecular imaging agents: impact on diagnosis and therapeutics in oncology. (2010) Expert Reviews in Molecular Medicine. 12: e20.

[2] Torigian DA, Huang SS, HouseiniM Functional imaging of cancer with emphasis on molecular techniques. *CA Cancer J Clin.* (2007) 57:206-224.

[3] Mankoff DA. A Definition of Molecular Imaging. *The Journal of Nuclear Medicine* (2007) 48(6): 18N-21N.

[4] Dunn AK, Bolay T, Moskowitz MA, Boas, DA Dynamic imaging of cerebral blood flow using laser speckle. *J Cereb Blood Flow Metab*(2001) 2:195-201.

[5] Alavi A, Lakhani P, Mavi A et al PET: a revolution in medical Imaging.*RadiolClin North* Am (2004) 42:983-1001.

[6] Antoch, G et al. Accuracy of whole-body dual-modality fluorine-18–2-fluoro-2-deoxy-D- glucose positron emission tomography and computed tomography (FDG-PET/CT) for tumor staging in solid tumors: comparison with CT and PET. *Journal of Clinical Oncology* (2004) 22, 4357-4368.

[7] Donati, OF et al. 18F-FDG-PET and MRI in patients with malignancies of the liver and pancreas. Accuracy of retrospective multimodality image registration by using the CT-component of PET/CT. *Nuklearmedizin* (2010) 49, 106-114.

[8] Barwick, T et al. Single photon emission computed tomography (SPECT)/computed tomography using iodine-123 in patients with differentiated thyroid cancer: additional value over whole body planar imaging and SPECT. *European Journal of Endocrinology* (2010) 162, 1131-1139.

[9] Mishra, A et al. A new class of Gd-based DO3A-ethylamine-derived targeted contrast agents for MR and optical imaging. Bioconjugate Chemistry (2006) 17, 773-780.

[10] Huber, MM et al. *Fluorescently detectable magnetic resonance imaging agents. Bioconjugate Chemistry* (1998) 9, 242-249.

[11] Brenner DJ et al. Computed Tomography — An Increasing Source of Radiation Exposure (2007) *N Engl J Med* 357(22): 2277-2284.

[12] Sampath L, Wang W, Sevick-Muraca EM) Near infrared fluorescent optical imaging for nodal staging. *J Biomed Opt* (2008) 13:041312.

[13] Hoffman RM Imaging tumor angiogenesis with fluorescent proteins. *APMIS* (2004) 112:441-449.

[14] Prasher DC, Eckenrodeb VK, Wardc WW et al. Primary structure of the Aequorea victoria green-fluorescent protein.(1992) *Gene* 111:229-233.

[15] Shaner NC, Steinbach PA, Tsien RY A guide to choosing fluorescent proteins. *Nature* (2005) 2:905-909.

[16] Ntziachristos V, Bremer C, Weissleder R. Fluorescence imaging with near-infrared light: new technological advances that enable in vivo molecular imaging. *EurRadiol*(2003) 13:195-208.

[17] Weissleder R, Ntziachristos V. Shedding light onto live molecular targets. Nat Medicine (2003) 9:123-128.

[18] Gould SJ and Subramani S Firefly luciferase as a tool in molecular and cell biology.*Anal Biochem*(1988) 175:5-13.

[19] Hastings JW. Biological diversity, chemical mechanisms, and the evolutionary origins of bioluminescent systems. *J MolEvol* (1983) 19: 309-321.

[20] Edinger M et al. Advancing animal models of neoplasia through in vivo bioluminescence imaging. *European Journal of Cancer* (2002) 38: 2128–2136.

[21] Pfleger KDG, Eidne K. Illuminating insights into protein-protein interactions using bioluminescence resonance energy transfer (BRET). *Nat Methods* (2006) 3:165-173.

[22] Michelini E, Mirasoli M, Karp M et al. Development of a Bioluminescence Resonance Energy-Transfer Assay for Estrogen-Like Compound *in Vivo* Monitoring. *Anal Chem* (2004) 76: 7069-7076 .

[23] Prinz A, Diskar M, Herberg FW Application of bioluminescence resonance energy transfer (BRET) for biomolecular interaction studies.*Chembiochem* (2006) 7:1007-1012.

[24] So MK, Xu C, Loening AM Self-illuminating quantum dot conjugates for in vivo imaging. *Nat Biotechnol* (2006) 24:339-343.

[25] Medintz IL et al. Quantum dotbioconjugatesforimaging, labelling and sensing. *Nat Mat* (2005) 4(6):435-46.

[26] Keereweer S et al. Optical Image-guided Surgery—Where Do We Stand? *Mol Imaging Biol (*2010) [Epub ahead of print].

[27] Wunderbaldinger et al. Near-infrared fluorescence imaging of lymph nodes using a new enzyme sensing activatable macromolecular optical probe. *EurRadiol* (2003) 13:2206–2211.

[28] Mahmood U et al. Near-Infrared Optical Imaging of Proteases in Cancer (2003) Mol*Cancer Ther*2: 489 – 496.

[29] Soltesz EG, Kim S, Kim SW et al. Sentinel lymph node mapping of the gastrointestinal tract by using invisible light. *Ann SurgOncol* (2006) 13:386-96.

[30] Tanaka E, Choi HS, Fujii H et al. Image-guided oncologic surgery using invisible light: Completed pre-clinical development for sentinel lymph node mapping. *Ann SurgOnc* (2006) 13:1671-81.

[31] Alberts, B et al. *Molecular biology of thecell.* 5th Ed. New York: Garland Science; 2008.

[32] Maxwell PH, Wiesener MS, Chang GW et al (1999) The tumour suppressor protein VHL targets hypoxia-inducible factors for oxygen-dependent proteolysis. *Nature*399:271-275.

[33] Bárdos JI, Ashcroft M Negative and positive regulation of HIF-1: a complex network. *BioEssays* (2004) 26:262-269.
[34] Gordan JD, Simon MC Hypoxia-inducible factors: central regulators of the tumor phenotype. *CurrOpin Genet Dev* (2007) 17:71-77.
[35] Evans SM, Judy KD, Dunphy I et al. Hypoxia is important in the biology and aggression of human glial brain tumors. *Clin Cancer Res* (2004) 10:8177-8184 .
[36] Furlan D. Sahnane N, Carnevali I et al. Regulation and stabilization of HIF-1alpha in colorectal carcinomas. *SurgOncol* (2007) 16:S25-S27.
[37] Victor N, Ivy A, Jiang BH, Agani FH Involvement of HIF-1 in invasion of Mum2B uveal melanoma cells. *ClinExp Metastasis* (2006) 23:87-96.
[38] Salnikow K, et al. *The Regulation of Hypoxic Genes by Calcium Involves c-Jun/AP-1, which Cooperates with Hypoxia-Inducible Factor 1 in Response to Hypoxia* (2002) 22(6): 1734–1741.
[39] Akagi,Y et al. Regulation of vascular endothelial growth factor expression in human colon cancer by insulin-like growth factor-I. *Cancer Res* (1998)58: 4008-4014.
[40] Iglesias P, Costoya JA. A novel BRET-based genetically encoded biosensor for functional imaging of hypoxia. *BiosensBioelec* (2009) 10:3126-30.
[41] Lee JY, Lee YS, Kim KL et al. A novel chimeric promoter that is highly responsive to hypoxia and metals.*Gene Therapy* (2006) 13:857-868.
[42] Shaner NC, Campbell RE, Steinbach PA et al. Improved monomeric red, orange and yellow fluorescent proteins derived from Discosoma sp. red fluorescent protein.*Nat Biotechnol* (2004) 22:1567-1572.
[43] Baird GS, Zacharias DA, Tsien RY Biochemistry, mutagenesis, and oligomerization of DsRed, a red fluorescent protein from coral. *ProcNatlAcadSci USA* (2000) 97:11984-11989.
[44] Ai H, Shaner NC, Cheng Z et al. Exploration of New Chromophore Structures Leads to the Identification of Improved Blue Fluorescent Proteins. *Biochemistry* (2007) 46:5904-5910.
[45] Shu X, Shaner NC, Yarbrough CA et al. Novel chromophores and buried charges control color in mFruits. *Biochemistry* (2006) 45:9639-9647.
[46] Hoffman RM Recent advances on in vivo imaging with fluorescent proteins. *Nat Rev Cancer* (2005) 5:796-806.
[47] Safran M, Kim WY, O'Connell F et al. Mouse model for noninvasive imaging of HIF prolyl hydroxylase activity: assessment of an oral agent that stimulates erythropoietin production. *ProcNatlAcadSci USA*. (2006) 103:105-110.
[48] Dinca EB et al. Bioluminescence imaging of invasive intracranial xenografts: implications for translational research and targeted therapeutics of brain tumors *Neurosurg Rev* (2010) 33:385–394.

In: Bioluminescence
Editor: David J. Rodgerson, pp. 129-136
ISBN 978-1-61209-747-3

Chapter 7

KINETICALLY MONITORING THE EXPRESSION OF ALPHA-TOXIN USING AN *HLA* PROMOTER-LUCIFERASE REPORTER FUSION IN *STAPHYLOCOCCUS AUREUS*

Yinduo Ji[*] and Jeffrey W. Hall
Department of Veterinary and Biomedical Sciences,
University of Minnesota, Minneapolois, MN 55455, USA

ABSTRACT

Staphylocoocus aureus is an important pathogen and can cause both human and animal infections. The continual emergence of drug resistant, notably methicillin resistant *S. aureus* isolates from hospitalized patients and in the community, as well as in livestock farms has produced a serious public health burden. The availability of a sensitive approach to monitor temporal gene expression or regulation enables us to elucidate molecular mechanism of resistance and pathogenesis. In this study, we chose bacterial luciferase (LuxABCDE) as a reporter and created a promoter-*lux* fusion in different *S. aureus* isolates, determined the transcriptional levels of alpha-toxin gene (*hla)* using a luminometer, and found that the temporal expression and the intensity of bioluminescence vary in different isolates carrying the same *hla* promoter-*lux* reporter. The results indicate that the luciferase-driven bioluminescence reporter provides a powerful tool for temporal examination of gene expression and regulation.

STAPHYLOCOCCAL ALPHA-TOXIN

The continuing emergence of hospital- and community-associated methicillin resistant *Staphylococcus aureus* infections highlights an urgent need for alternative potent antibacterial

[*] Corresponding author: 1971 Commonwealth Ave., St. Paul, MN 55108, Telephone: 612-624-2757, Fax: 612-625-5203, E-mail: jixxx002@umn.edu

agents. The ability of this organism to resist antibiotics and cause infection is partially due to the coordinated regulation of gene expression, which allows the bacteria to survive in different stress conditions. Two-component signal (TCS) transduction systems play important roles in the adaptation of the microbial organisms within different niches, as well as in pathogenesis and biofilm formation for various bacterial species [1-4].

Alpha-toxin is an important virulence factor in experimental brain abscesses [5], pneumonia [6] and intraperitoneal infections [7,8]. Moreover, α-toxin plays a critical role in *S. aureus* mastitis [9,10]. It has been found that significant increases in milk antibodies to α- and β-toxins are present in cows with chronic staphylococcal mastitis [11], and that many *S. aureus* isolates from the mammary gland of dairy cows produce α-toxin [12] and some bovine *S. aureus* isolates generate dramatic amounts of α-toxin [13]. However, the role played by α-toxin varies according to the stage of infection and the quantities produced. It has been revealed that small colony variants (SCVs) have dramatically reduced production of α-toxin and consequently SVCs have an increased capacity for adhesion to human epithelial cells and persistence in infections [14, 15]. The overproduction of α-toxin significantly reduces virulence in an experimental endocarditis [16]. Taken together, the above data indicate that α-toxin likely plays different roles in pathogenesis during different stages of infection.

Alpha-toxin can specifically interact with the surface receptors of host cells, form functional transmembrane pores and selectively release ions, and/or trigger cell signal transduction pathways, thus inducing apoptosis and/or necrosis in various cell types [17-20]. Also, α-toxin can activate the host cell ectodomain shedding machinery, leading to the release of host cell receptors for ECM molecules, thus promoting bacterial pathogenesis [21]. We have demonstrated that α-toxin interacts with β1-integrin of epithelial cells, interrupts the binding of Fn to its receptor, β1-integrin, and blocks the formation of the bridge between the microbe and the host cells, which in turn inhibits adherence and internalization [22]. In addition, the interaction of α-toxin and β1-integrin contributes to the toxicity of α-toxin [23]. Our results suggest that *S. aureus* requires a lower level of α-toxin in order to initiate and establish infection; at a later stage of infection, a higher production of α-toxin and the subsequent inhibitory effect on adhesion may be required for persistent infection. The destruction of infected host cells and inhibition of adherence to the damaged cells may be an advantageous process for penetration into new ecological niches.

Bioluminescence Reporter

Luciferases have been used widely to study gene expression and protein localization in eukaryotic cells because light is easily detectable and can be quantified with high precision. Bioluminescence also has the advantage that no external light source is required to excite the reporter. Luciferase refers to a broad and evolutionarily diverse group of enzymes derived from insects, coelenterates, bacteria, and other organisms. The luciferases and the subsequent reactions from these diverse organisms are different. Beetle luciferases from the common firefly, *Photinus pyralis*, or the click beetle, *Pyrophorus plagiophthalamus*, use beetle luciferin, oxygen and adenosine 5'-triphosphate (ATP) as the substrates; Coelenterates, such as the sea pansy (*Renilla reniformis*), require coelenterazine and oxygen only as the substrates [24]. However, in bioluminescent bacteria such as *Photorhabdus luminescens* or *Vibrio harveyi*, light emission results from an oxidative reaction of flavin mononucleotide (FMNH2) and a long-chain aliphatic aldehyde [25, 26]. The beetle luciferases and Renilla luciferase are synthesized as active monomeric enzymes when expressed in bacteria and require no post-translational modifications [24]. The bacterial *lux* system is completely different from the other luciferase systems and when used as a reporter in bacterial cells, the entire *lux* operon consisting of five genes, *luxCDABE*, must be expressed, and the detection sensitivity is reduced. However, expression from the entire operon confers the advantage that no exogenous substrates need to be added.

The *lux* system has been extensively used to monitor gene expression both *in vitro* culture and *in vivo* during *S. aureus* infection [27-29], as well as a whole-cell bacterial sensor [24-26]. With development of highly sensitive instruments to detect and measure light production it is now possible to evaluate persistence of bacteria in live rodents. Bioluminescence has been successfully used to monitor the persistence of group A streptococcus (GAS) and *S. aureus* using different rodent models of infection [30, 31, 28]. In these studies the *luxABCDE* Km^+ cassette was inserted into the chromosome of GAS or located in a plasmid. The resulting recombinant produced visible light only when growing and expanding their numbers. Mice were infected, anesthetized at various times after infection and placed in a dark chamber of Xenogen $IVIS^{100}$ luminometer. Photons could be collected and quantified by the instrument over a period of time. Because the infected animals are not killed in order to evaluate bacterial persistence and expansion, this system requires considerably fewer animals and expedites the analysis.

Kinetically Monitoring the Expression of Alpha-Toxin Using a *hla* Promoter-*Lux* Reporter Fusion

Different promoter-*lacZ* and promoter-*gfp* fusion systems have been widely utilized for monitoring gene expression by measuring LacZ and Gfp production in prokaryotic cells [32, 33]. However, the long half-time of degradation and high background produced by these two proteins limit their usage in the investigation of temporal gene expression and regulation. It has been reported that a modified *luxABCDE* (*lux*) cassette can be used for transient expression in $Gram^+$ bacteria [27]. We chose *lux* as a reporter because of its high sensitivity and good correlation of the luminescent signal with viable bacterial counts relative to current fluorescent approaches.

Alpha-toxin is highly expressed during the stationary growth phase *in vitro* [34] and at later stages of infection in the animal [35]. The regulation of *hla* expression has been extensively studied; it has been revealed that the *hla* gene is tightly regulated by different regulators, including TCS, such as the accessory gene regulator (Agr) [34, 36], the staphylococcal accessory protein effector (SaeRS) [36, 37] and ArlRS [36, 38], and transcriptional regulators Mgr [39]) and SarZ [40]. The staphylococcal accessory regulator (SarA) positively affects *hla* expression in both *agr*-dependent and *agr*-independent manners [41, 34], whereas the Rot and SarT, which are homologues of SarA, repress the expression of α-toxin [42, 43]. The alternative sigma B factor (σ^B) and environmental stimuli also have an influence on *hla* transcription [44]. Collectively, the above data indicate that kinetically monitoring of gene expression is important for us to elucidate the mechanism of gene regulation and pathogenesis of *S. aureus*. We created a *hla* promoter-*lux* reporter fusion and introduced it into four *S. aureus* isolates, including the laboratory strain RN4220 and its derivative, RN6390, as well as the human clinical isolate, WCUH29, and the cow mastitis isolate, RF122, to monitor the activity of *hla* promoter in real time during bacterial growth in culture.

We found that the overall pattern of *hla* promoter-*lux* reporter expression was similar in 3 of the 4 strains, with maximal expression of the *hla* promoter-*lux* reporter occurring from mid-log to the late exponential phase of growth, but the amount of light detected varied greatly between the strains (Figure 1A-C). Interestingly, bioluminescence was essentially undetectable from the RF122 strain (Figure 1D). *S. aureus* RN4220 is a highly mutagenized laboratory strain and gene expression within this strain may not accurately reflect the actual level of gene expression in other strains; this is exemplified by the 40- and 10-fold increase of reporter expression compared to the RN6390 and WCUH29 strains, respectively (Fig. 1A-C). Moreover, transcription from the *hla* promoter varied during the early phase of growth among the strains, with RN4220 showing increasing expression to late exponential phase (Fig. 1A), RN6390 showed a start and stop like expression early in growth, with a continual increase of expression until late exponential (Figure 1B). The WCUH29 strain showed enhanced *hla* transcription earlier in the growth cycle and maximal reporter expression occurred at the mid-log phase of growth. The bacterial strains grew at approximately the same rate, with some cell death after extended periods of incubation (Figure 1).

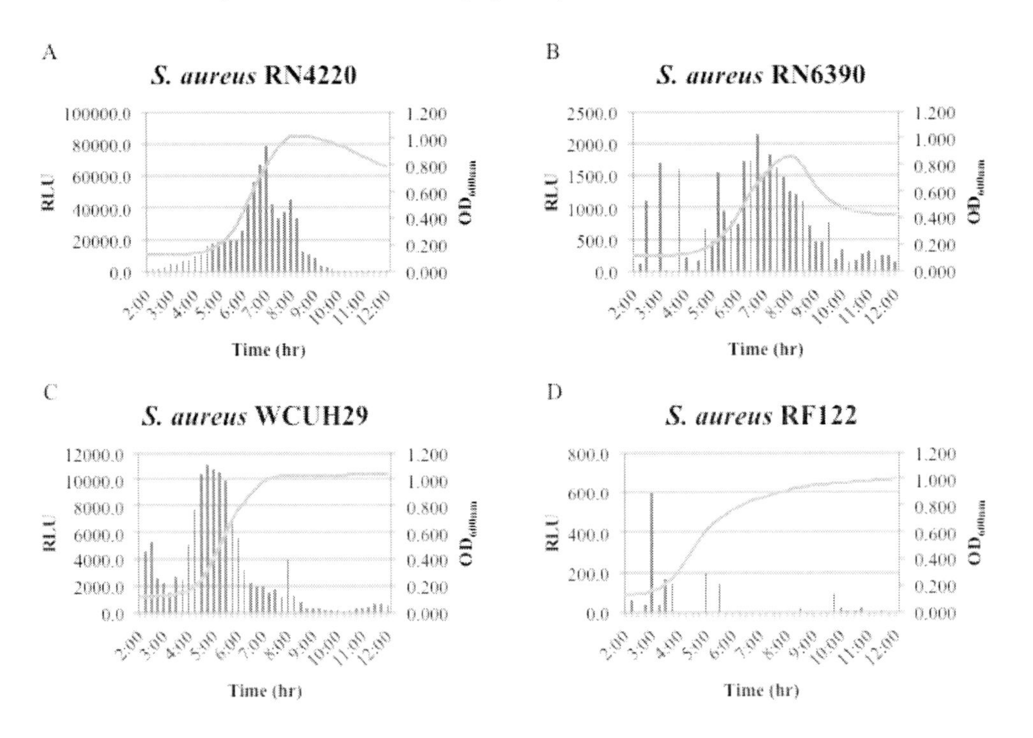

Figure 1. *hla* promoter-*lux* reporter bioluminescence from four *S. aureus* strains. Left vertical axis, column graph, is the relative bioluminescence value (Light Value/ OD_{600nm} = RLU) for each time point. Right vertical axis, line graph, is the optical density at 600 nm for each time point. (A) Laboratory strain, RN4220. (B) Laboratory strain, RN6390. (C) Human clinical isolate, WCUH29. (D) Bovine mastitis isolate, RF122.

The inability to detect light from the RF122 is not due to a failure of the reporter system [Liang et al., unpublished data], as the same plasmid was introduced into all four strains, or a failure to detect light by the luminometer, but is likely the result of an intrinsic characteristic of the RF122 bacterial cell. The strains were analyzed using a BioTek Synergy 2 in a 96-well format, with the bioluminescence and optical density at 600nm being recorded at each time point, and across several light sensitivity settings (data not shown). Bioluminescence production that would be consistent with the known *hla* transcription pattern was not detected from the RF122 strain during any of the experiments (Figure 1D). Figure 1 represents data obtained from experiments with the light detection setting at almost maximal sensitivity.

ACKNOWLEDGMENTS

We thank Dr. Xudong Liang and Ms. Junshu Yang for technical assistance. This study was supported by the MAES Competitive Grants and partially by NIH grant AI057451.

REFERENCES

[1] Gross, R. (1993). Signal transduction and virulence regulation in human and animal pathogens. *FEMS Microbiol.* 10, 301.

[2] Stock, A. M., Robinson, V. L., & Goudreau, P. N. (2000). Two-component signal transduction. *Annu. Rev. Biochem.*, 69,183.

[3] Li, Y. H., Lau, P.C.Y., Tang, N., Svensater, G., Ellen, R. P., and Cvitkovitch, D. G. (2002). Novel two-component regulatory system involved in biofilm formation and acid resistance in Streptococcus mutans. *J. Bacterol.* 184,6333.

[4] Novick, R. (2006). Staphylococcal pathogenesis and pathogenicity factors: genetics and regulation, p. 496-516. In Fischetti et al. (ed.), gram-positive pathogens. *ASM Press*, Washington, D.C.

[5] Kielian, T., Cheung, A., and Hickey, W. F. (2001). Diminished Virulence of an Alpha-Toxin Mutant of Staphylococcus aureus in Experimental Brain Abscesses. *Infect. Immun.* 69,6902.

[6] Wardenburg, J. B., Patel, R. J., and Schneewind, O. (2007). Surface proteins and exotoxins are required for the pathogenesis of Staphylococcus aureus pneumonia. *Infect. Immun.* 75,1040.

[7] Kernodle, D. S., Voladri, R. K., Menzies, B. E., Hager, C. C., and Edwards, K. M. (1997). Expression of an antisense hla fragment in Staphylococcus aureus reduces alpha-toxin production in vitro and attenuates lethal activity in murine model. *Infect. Immun.* 65,179.

[8] Ji, Y., Marra, A., Rosenberg, M., and Woodnutt, G. (1999). Regulated Antisense RNA Eliminates Alpha-Toxin Virulence in Staphylococcus aureus Infection. *J. Bacteriol.* 181,6585.

[9] Bramley, A. J.,, Patel, A. H., O'Reilly, M., Foster, R., and Foster, T. J. (1989). Roles of alpha-toxin and beta-toxin in virulence of Staphylococcus aureus for the mouse mammary gland. *Infect. Immun.* 57,2489.

[10] Jonsson, P., Lindberg, M., Haraldsson, I., and Wadström, T. (1985). Virulence of Staphylococcus aureus in a mouse mastitis model: studies of alpha hemolysin, coagulase, and protein A as possible virulence determinants with protoplast fusion and gene cloning. *Infect. Immun.* 49,765.

[11] Loeffler, D.A., and Norcross, N. L. (1987). Use of enzyme-linked immunosorbent assay to measure bovine milk and serum antibodies to alpha toxin, beta toxin, and capsular antigens of Staphylococcus aureus. *Vet. Immunol. Immunopathol.* 14,145.

[12] Kenny, K., Bastida, F. D., and Norcross, N. L. (1992). Secretion of alpha-hemolysin by bovine mammary isolates of Staphylococcus aureus. *Can. J. Vet. Res.* 56,265.

[13] Guinane, C. M., Sturdevant, D. E., Herron-Olson, L., Otto, M., Smyth, D. S., et al. (2008). Pathogenomic analysis of the common bovine Staphylococcus aureus clone (ET3): emergence of a virulent subtype with potential risk to public health. *J. Infect. Dis.* 197,205.

[14] Sadowska, B., Bonar, A., von Eiff, C., Proctor, R.A., Chmiela, M., Rudnicka, W., and Rozalska, B. (2002). Characteristics of Staphylococcus aureus, isolated from airways of cystic fibrosis patients, and their small colony variants. *FEMS Immunol. Med. Microbiol.* 32,191.

[15] Proctor, P.A., Langevelde, P., Kristjansson, M., Maslow, J.N., and Arbeit, D.D. (1995). Persistent and relapsing infections associated with small-colony variants of Staphylococcus aureus. *Clin. Infect. Dis.* 20,95.

[16] Bayer, A., Ramos, M., Menzies, B., Yeaman, M., Shen, A., et al. (1997). Hyperproduction of alpha-toxin by Staphylococcus aureus results in paradoxically reduced virulence in experimental endocarditis: a host defense role for platelet microbicidal proteins. *Infect. Immun.* 65,4652.

[17] Song, L., Hobaugh, M. R., Shustak, C., Cheley, S., Bayley, H., et al. (1996). Structure of staphylococcal α-hemolysin, a heptameric transmembrane pore. *Science* 274,1859.

[18] Menzies, B. E., and Kourteva, I. (2000). Staphylococcus aureus α-toxin induces apoptosis in endothelial cells. *FEMS Immunol. Med. Microbiol.* 29,39.

[19] Essmann, F., Bantel, H., Totzke, G., Engels, I. H., Sinha, B., et al. (2003). Staphylococcus aureus alpha-toxin-induced cell death: predominant necrosis despite apoptotic caspase activation. *Cell Death Differ.* 10,1260.

[20] Haslinger, B., Strangfeld, K., Peters, G., Schulze-osthoff, K., and Sinha, B. (2003) Staphylococcus aureus α-toxin induces apoptosis in peripheral blood mononuclear cells: role of endogenous tumor necrosis factor-α and the mitochondrial death pathway. *Cell. Microbiol.* 5,729.

[21] Park, P., Foster, T., Nishi, E., Duncan, S., Klagsbrun, M., and Chen, Y. (2004). Activation of syndecan-1 ectodomain shedding by Staphylococcus aureus alpha-toxin and beta-toxin. *J. Biol. Chem.* 279,251.

[22] Liang, X., and Ji, Y. (2006). Alpha-toxin interferes with integrin-mediated adhesion and internalization of Staphylococcus aureus by epithelial cells. *Cell. Microbiol.* 8,1656.

[23] Liang, X., and Ji, Y. (2007). Involvement of alpha5beta1-integrin and TNF-alpha in Staphylococcus aureus alpha-toxin-induced death of epithelial cells. *Cell. Microbiol.* 9,1809.

[24] Wilson, T., and Hastings, J. W. (1998). Bioluminescence. *Annu. Rev. Cell. Dev. Biol.* 14,197.

[25] Wood, K. V. (1995). The chemical mechanism and evolutionary development of beetle bioluminescence. *Photochem. Photobiol.* 62,662.

[26] Fan, F., Binkowski, B. F., Butler, B. L., Stecha, P. F., Lewis, M. K., and Wood, K.V. (2008). Novel Genetically Encoded Biosensors Using Firefly Luciferase. *ACS Chemical Biol.* 3,346.

[27] Qazi, S. N., Harrison, S. E., Self, T., Williams, P., and Hill, P. J. (2004). Real-time monitoring of intracellular Staphylococcus aureus replication. *J. Bacteriol.* 186,1065.

[28] Wright, J., Jin, R., and Novick, R. P. (2005). Transient interference with staphylococcal quorum sensing blocks abscess formation. Proc. *Natl. Acad. Sci. USA* 102,1691.

[29] Yan, M., Yu, C., Yang, J., and Ji, Y. (2009). Development of shuttle vectors for evaluation of essential gene regulation in Staphylococcus aureus. *Plasmid* 61,188.

[30] Francis, K. P., Yu, J., Bellinger-Kawahara, C., Joh, D., Hawkinson, M.J., Xiao, G., Purchio, T. F., Caparon, M. G., Lipsitch, M., and Contag, P. R. (2001). Visualizing pneumococcal infections in the lungs of live mice using bioluminescent Streptococcus pneumoniae transformed with a novel Gram-positive lux transposon. *Infect. Immun.* 69,3350.

[31] Park, H., Francis, K., Yu, J., and Cleary, P. (2003). Membranous cells in nasal-associated lymphoid tissue: a portal of entry for the respiratory mucosal pathogen group A streptococcus. *J. Immunol.* 171,2532.

[32] Ohlsen, K., Koller, K. P., and Hacker, J. (1997). Analysis of expression of the alpha-toxin gene (hla) of Staphylococcus aureus by using a chromosomally encoded hla::lacZ gene fusion. *Infect. Immun.* 65,3606.

[33] Qazi, S., Rees, C., Mellits, K., and Hill, P. J. (2001). Development of gfp vectors for expression in Listeria monocytogenes and other low G+C gram-positive bacteria. *Microb. Ecol.* 41,301.

[34] Novick, R.P. (2003), Autoinduction and signal transduction in the regulation of staphylococcal virulence. *Mol. Microbiol.* 48,1429.

[35] Da Silva, M., Zahm, J., Gras, D., Bjolet, O., Abely, M., et al. (2004). Dynamic interaction between airway epithelial cells and Staphylococcus aureus. *Am. J. Physiol. Lung Cell Mol. Physiol.* 287,L453.

[36] Goerke, C., Fluckiger, U., Steinhuber, A., Zimmerli, W., and Wolz, C. (2001). Impact of the regulatory loci agr, sarA and sae of Staphylococcus aureus on the induction of α-toxin during device-related infection resolved by direct quantitative transcript analysis. *Mol. Microbiol.* 40,1439.

[37] Liang, X., Yu, C., Sun, J., Liu, H., Landwehr, C., Holmes, D., and Ji, Y. (2006). Inactivation of a two-component signal transduction system, SaeRS, eliminates adherence and attenuates virulence of Staphylococcus aureus. *Infect. Immun.* 74,4655.

[38] Liang, X., Zheng, L., Landwehr, C., Lunsford, D., Holmes, D., and Ji, Y. (2005). Global regulation of gene expression by ArlRS, a two-component signal transduction regulatory system of Staphylococcus aureus. *J. Bacteriol.* 187,5486.

[39] Luong, T.T., Newell, S. W., and Lee, C. Y. (2003). Mgr, a novel global regulator in Staphylococcus aureus. *J. Bacteriol.* 185,3703.

[40] Ballal, A., Ray, B., and Manna, A. C. (2009). sarZ, a sarA Family Gene, Is Transcriptionally Activated by MgrA and Is Involved in the Regulation of Genes Encoding Exoproteins in Staphylococcus aureus. *J. Bacteriol.* 191,1656.

[41] Oscarsson, J., Kanth, A., Tegmark-Wisell, K., and Arvidson, S. (2006). SarA is a repressor of hla (alpha-hemolysin) transcription in Staphylococcus aureus: its apparent role as an activator of hla in the prototype strain NCTC 8325 depends on reduced expression of sarS. *J. Bacteriol.* 188,8526.

[42] McNamara, P. J., Milligan-Monroe, K. C., Khalili, S., and Proctor, R. A. (2000). Identification, cloning, and initial characterization of rot, a locus encoding a regulator of virulence factor expression in Staphylococcus aureus. *J. Bacteriol.* 182,3197.

[43] Schmidt, K. A., Manna, A. C., Gill, S., and Cheung, A. (2001). SarT, a repressor of alpha-hemolysin in Staphylococcus aureus. *Infect. Immun.* 69,4749.

[44] Karlsson-Kanth, A., Tegmark-Wisell, K., Arvidson, S., and Oscarsson, J. (2006) Natural human isolates of Staphylococcus aureus selected for high production of proteases and alpha-hemolysin are sigmaB deficient. *Int .J. Med. Microbiol.* 296,229.

In: Bioluminescence
Editor: David J. Rodgerson, pp. 137-151

ISBN 978-1-61209-747-3

Chapter 8

BRET-BASED SCREENING ASSAYS IN SEVEN-TRANSMEMBRANE (7TM) RECEPTOR DRUG DISCOVERY

***Anders Heding*[1] *and Milka Vrecl*[2]**
[1]Protein Chemistry Biopharm, Novo Nordisk A/S, Måløv, Denmark
[2]Institute of Anatomy, Histology & Embryology, Veterinary Faculty, University of Ljubljana, Ljubljana, Slovenia

ABSTRACT

Bioluminescence resonance energy transfer (BRET) represents a biophysical method to study physical interactions between protein partners in living cells fused to donor and acceptor moieties. It relies on a non-radiative transfer of energy between donor and acceptor, their intermolecular distance (10 – 100 Å) and relative orientation. Several versions of BRET have been developed that use different substrates and/or energy donor/acceptor couples to improve stability and specificity of the BRET signal. In recent years, numerous studies have applied BRET technology to develop screening assays for seven-transmembrane receptors (7TMRs), which represent a key drug target class. In general, these assays are based on 7TMR/β-arrestin interaction, common to virtually all 7TMRs; however, differences in 7TMRs affinity for β-arrestins and the stability/longevity of receptor/β-arrestin complexes exist. This chapter summarizes different approaches (e.g. mutations in β-arrestins, 7TM receptor carboxyl-terminal tail swapping) to optimize the BRET assay for measuring receptor/β-arrestin interaction and applicability of this technological platform for compound medium/high-throughput screening (MHT/HTS).

Keywords: Bioluminescence resonance energy transfer; 7TM receptors; beta-arrestins; protein-protein interaction; screening

INTRODUCTION

Seven transmembrane receptors (7TMRs; G-protein coupled receptors (GPCRs)) form the largest and evolutionarily well-conserved family of cell-surface receptors, with more than 800 members identified in the human genome [1]. They play a fundamental role in the regulation of many physiological processes in the organism. Consequently, disturbances in their action are associated with the pathophysiology of numerous diseases such as nephrogenic diabetes insipidus, *reproductive system disorders* and cardiovascular, respiratory, central and peripheral nervous system diseases. At present, 7TMRs are one of the most important drug targets in the pharmaceutical industry; approximately 40% of the prescription drugs in the market target 7TMRs, but only 5% of the known 7TMR targets are utilized [2]. Therefore, this receptor family remains a central focus in basic pharmacology studies and drug discovery efforts.

An essential step in the initial phase of drug discovery is the generation of hits, usually by screening large chemical compound libraries against the target (receptor). For the past two decades, binding, Ca^{2+} transient assays and GTPγS (guanosine-5'-O-(3'-thiotriphosphate)) assays dominated among screening assays in the 7TMRs field [3]. Binding assays were based on the displacement of a radiolabeled ligand from the receptor via addition of compound. These assays cannot differentiate between agonist and antagonists, usually run in an unnatural environment as they utilize purified or semipurified receptors, i.e. membrane fragments attached to a surface (e.g. bead or microplate) and consequently result in many false hits [4]. Therefore, binding assays were gradually replaced by cell-based functional assays that monitor functional responses elicited by 7TMR activation. The canonical model for 7TMR signal transduction is comprised of a receptor, a heterotrimeric G-protein and an effector protein. Inherent to all 7TMRs is their ability to bind to and activate heterotrimeric G-proteins, which transduce receptor signal across the plasma membrane. Thus, functional cell-based screening on 7TMRs is based on the activation of a specific signal transduction pathway via coupling to three main classes of G-proteins, i.e. $G\alpha_s$, $G\alpha_{i/0}$, or $G\alpha_{q/11}$. The activation of $G\alpha_s$- and $G\alpha_i$-coupled 7TMRs is commonly monitored by measurements (increase/decrease) of intracellular cAMP levels, while the activation of $G\alpha_{q/11}$-coupled 7TMRs can be monitored either by intracellular inositol phosphate (IP) or Ca^{2+} accumulation [5]. Due to their limited detection range from the outset to a particular signaling pathway, distinct assay platforms are needed to screen 7TMRs in a functional cell-based way. This is both time and resource demanding, as different assays/methods as well as different instruments have to be acquired and established [6]. Therefore, there has long been a wish for a universally applicable, functional, cell-based screening assay, which could be used for the screening of all receptors independent of their signaling pathway. In the classical assay systems, in which ligand-activation of 7TMRs is translated into G-protein-dependent intracellular responses, artificially created promiscuous G-proteins could fulfill this criteria as they interact with a wide range of 7TMRs and can be utilized to force some 7TMRs (especially $G\alpha_i$-coupled receptors) to signal through the $G\alpha_q$ pathway [7]. However, recent technological advances such as the development of non-invasive fluorescence-based detection and imaging techniques for determining protein-protein interaction [8] has greatly expanded the possible modes of detection of 7TMR function. A number of assays have been developed that do not rely on G-protein activation, but on an almost ubiquitous interaction of 7TMRs with β-arrestins.

7TMRs and Arrestins

The arrestin family has four members in mammals (i) arrestin1 (called visual or rod arrestin in some species, and previously called S-antigen or 48 kDa protein), (ii) arrestin2 (also known as β-arrestin or β-arrestin1), iii) arrestin3 (β-arrestin2) and (iv) arrestin4 (cone arrestin or X-arrestin). Arrestin1 and 4 form the subfamily of visual or sensory arrestins, while arestin2 (β-arrestin1) and 3 (β-arrestin2) form subfamily of non-visual arrestins. In contrast to highly specialized visual arrestins that are selectively expressed at very high levels in retinal photoreceptor cells and bind rhodopsin and cone opsins, the two non-visual arrestin subtypes are ubiquitously expressed in virtually every cell, and interact with the vast majority of 7TMRs independent of their signaling pathway, however, β-arrestin2 demonstrates higher affinity for many 7TMRs than β-arrestin1 (for review see [9]). Agonist binding to 7TMRs results in both the initiation of cellular signaling and the receptor phosphorylation by G protein-coupled receptor kinases (GRKs). β-arrestins binding to GRKs phosphorylated active 7TMR terminate G-protein mediated signaling (receptor desensitization) and target receptors to coated pits for internalization.

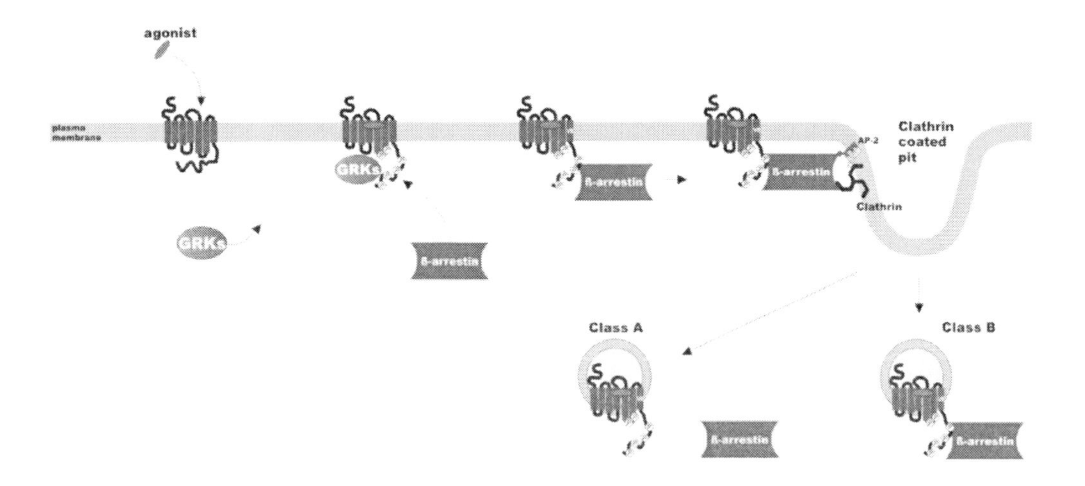

Figure 1. Schematic model of the β-arrestin-mediated Class A and B receptor internalization. After agonist binding, a 7TMR changes conformation to an active state, which leads to phosphorylation of the receptor by GRKs, generating a high affinity binding sites for β-arrs. Following binding to the receptor, β-arr targets the 7TMRs into the clathrin-coated pits via interaction with the coated pits components AP-2 and clathrin for subsequent internalization. The stability of the receptor/β-arr complex delineates two receptor classes (A and B). Class A receptors bind β-arrs with lower affinity than class B receptors and dissociate from β-arrs at the vicinity of the plasma membrane during/immediately after internalization. Class B receptors remain in complex with β-arrs for prolonged period of time after internalization. (From Vrecl et al., Journal of Biomolecular Screening (vol. 9, No. 4, pp 322-33). Copyright © 2004 by (SAGE Publications). Reprinted by Permission of SAGE Publications).

This is achieved by the interaction of 7TMR-bound β-arrestins with major components of coated pits such as clathrin and adaptor protein-2 (AP-2), albeit with different affinities. β-arrestin2 has higher affinity for clathrin and AP-2 compared to β-arrestin1 [10, 11].

Differences in receptor affinities for β-arrestin1 and β-arrestin2, and their ability to remain in complex with β-arrestins after internalization delineate two major receptor classes (Class A and B) [12] as schematically shown in Figure 1. After agonist stimulation, the Class A receptors (e.g. β_2-adrenergic receptor (β_2-AR), dopamine D1A receptor, and β_{1b}-adrenergic receptor) recruit β-arrestin2 more efficiently than β-arrestin1 but then rapidly dissociate from the β-arrestins just after internalization in the vicinity of the plasma membrane. In contrast, the Class B receptors including neurokinin type 1 (NK1-R), vasopressin type 2 (V2R), angiotensin II type 1A receptor, and type I thyrotropin-releasing hormone receptor recruit both β-arrestins equally well and remain bound to the β-arrestins for a prolonged time period after internalization [12]. However, some 7TMRs do not fit into this classification. For instance, the metabotropic glutamate receptor 1a interacts only with β-arrestin1 [13], somatostatin receptor 2A internalizes with β-arrestin1, yet is recycled and resensitized rapidly, somatostatin receptor 3 releases arrestin near the plasma membrane, but does not recycle rapidly to the plasma membrane [14] and a small number of receptors (e.g. somatostatin receptor 4, γ-aminobutyric acid type B ($GABA_B$) and mammalian type I gonadotropin-releasing hormone receptors) appear not to recruit β-arrestins [14-16]. Nonetheless, 7TMR/β-arrestin interaction/cytoplasmic translocation has been exploited to develop new assays platforms that are based either on direct measurement of receptor/β-arrestin green fluorescent protein (GFP) complexes redistribution [17, 18] (TransFluor® Assay; Molecular Devices), bioluminescence resonance energy transfer (BRET) [19-22], enzyme fragment complementation [23, 24] (PathHunter™ Enzyme Fragment Complementation; DiscoveRx) or protease-activated transcriptional reporter genes [25] (Tango™ GPCR Assay System; Invitrogen). This chapter focuses on the development and optimization of BRET-based screening assays.

BRET Principle and Technology

In recent years, biophysical methods based on fluorescence and bioluminescence resonance energy transfer (respectively, FRET and BRET) have been developed. These techniques enable monitoring of physical interactions between two proteins fused to FRET/BRET donor and acceptor moieties, respectively, dependent on their intermolecular distance (10 – 100 Å) and on relative orientation due to the dipole-dipole nature of the resonance energy transfer mechanism [26]. Although, FRET enables the visualization of protein interactions in living cells, the problems associated with FRET (e.g. high fluorescent background, autofluorescence, photobleaching) make BRET the technology of choice for several applications. BRET is a non-radiative energy transfer, occurring between a bioluminescent donor that emits light in the presence of its corresponding substrate and a complementary fluorescent acceptor, which absorbs light at a given wavelength and re-emits light at longer wavelengths. To fulfill the condition for energy transfer, the emission spectrum of the donor must overlap with the excitation spectrum of the acceptor molecule [26]. BRET occurs naturally in some marine species (e.g. in the sea pansy *Renilla reniformis*) and in 1999,

Xu et al. [27] utilized this approach to study dimerization of the bacterial Kai B clock protein [27]. Subsequently, in 2000, BRET was introduced in the 7TMR field demonstrating β_2-AR dimerization and agonist-promoted β_2-AR/β-arrestin2 complex formation [28]. Several versions of BRET assays have been developed that use different substrates and/or energy donor/acceptor couples. The original BRET[1] technology used the pairing of *Renilla luciferase* (Rluc) as the donor and yellow fluorescent protein (YFP) as the acceptor [27, 29]. The addition of coelenterazine h, the natural substrate of *Renilla luciferase* (Rluc), leads to a donor emission of blue light (peak at ~480 nm). When the YFP-tagged acceptor molecule, adapted to this emission wavelength, is in close proximity to the Rluc-tagged donor molecule, excitation of YFP occurs by resonance energy transfer resulting in an acceptor emission of green light (peak at ~530 nm). The substantial overlap in the emission spectra of Rluc and YFP acceptor emission (Stokes shift only ~50 nm) creates a significant problem that has been overcome in a second generation BRET assay (BRET[2]). In BRET[2] assays, *Renilla luciferase* (Rluc) is used as the donor, the green fluorescent protein (GFP) variant GFP[2] as the acceptor molecule (excitation ~400 nm, emission peak at 510 nm) and the proprietary coelenterazine DeepBlueC™ (also known as coelenterazine 400A) as a substrate. In the presence of DeepBlueC™, Rluc emits light peaking at 395 nm, a wavelength that excites GFP[2] resulting in the emission of green light at 510 nm. This modified BRET pair results in a broader Stokes shift of 115 nm, thus enabling superior separation of donor and acceptor peaks. However, the disadvantage of BRET[2], compared to BRET[1] is the 100-300 times lower intensity of emitted light and a very fast decay of emitted light [6]. BRET[2] sensitivity can be improved by the development of suitably sensitive instruments [6] and the use of Rluc mutants with improved quantum efficiency and/or stability (e.g. Rluc8 and Rluc-M) as a donor [30]. A third generation BRET assay (BRET[3]) has been developed recently that combines Rluc8 with the mutant red fluorescent protein (DsRed2) variant mOrange and the coelenterazine or EnduRen™ as a substrate [31]. EnduRen™ is a very stable coelenterazine analogue that enables luminescence measurement for at least 24 hours after substrate addition and was utilized in the extended BRET (eBRET) technology [32]. Therefore, in BRET[3], donor spectrum is the same as in BRET[1], and the red shifted mOrange acceptor signal (emission peak at 564 nm) improves spectral resolution to 85 nm, thereby reducing bleedthrough in the acceptor window. Improved spectral resolution and increased photon intensity allow imaging of protein-protein interactions from intact living cells to small living subjects. Due to some limitations of the BRET technique (e.g. low intensity of the light generated by the Rluc, use of fusion proteins, interference of some chemicals with the luciferase activity), it is only recently that the BRET technology has been used as a screening platform [19-22, 33] simply because the luminescence readers now may be equipped with on-line injectors and has obtained the necessary sensitivity. In addition, different approaches such as mutation in β-arrestin2 and C-terminal tail swapping were used to improve signal window in the BRET-based β-arrestin assays [20-22].

BRET[2] Signal - Class A *vs.* Class B 7TMRs

Initial BRET experiments with 7TMRs C-terminally tagged with Rluc and human β-arrestin2 constructs N-terminally tagged with YFP/GFP[2] revealed that assay window for the

Class A 7TMRs was rather low, while agonist activation of Class B receptors resulted in a robust BRET signal [20, 21]. This may occur to such an extent that compound screening becomes impractical. An example of specific agonist-induced BRET^2 signals generated from Class A (β_2-AR) and Class B (NK1-R) interactions with wild type (WT) β-arrestin2 (β-arr2) can be seen in Figure 2. Here, the specific BRET^2 signal (the BRET^2 signal from cells expressing receptor/Rluc alone has been subtracted) from the Class B receptor is almost five-fold higher than the signal from the Class A receptor. To optimize BRET^2-based β-arrestin assay for the Class A 7TMRs β-arr2 mutants (7TM Pharma proprietary) have been created that do not interact with the components of clathrin-coated pits and therefore, form more stable and longer lasting β-arr2/receptor complexes. Thus, in principle, all receptors should behave as Class B 7TMRs. In order to understand the rationale behind this approach, a more detailed description of clathrin and AP-2 binding sites located in the *carboxyl-terminus* of the β-arrestins is given.

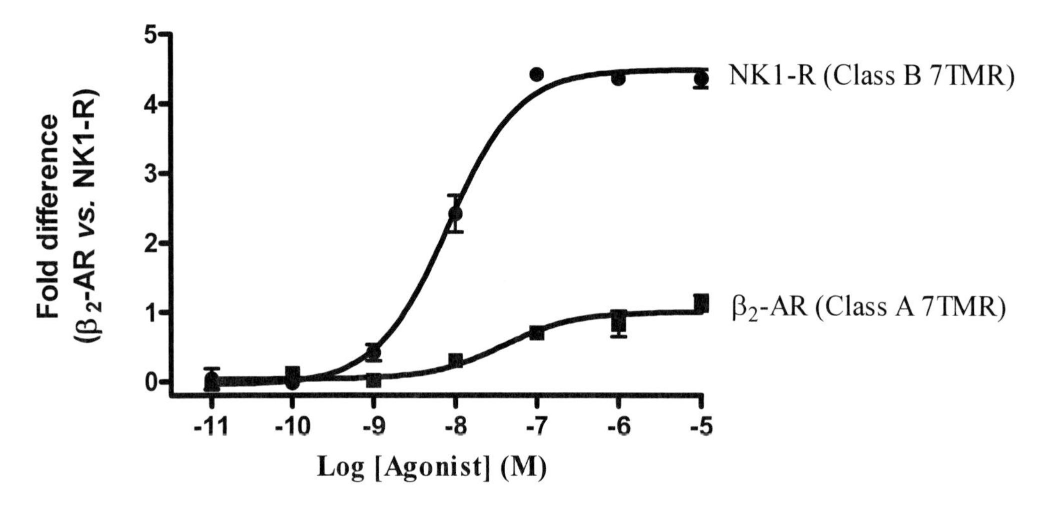

Figure 2. Differences between BRET^2 signals generated by agonist-induced GFP^2/β-arr2 interaction with Class A and Class B 7TMRs. Increasing concentrations of the adrenergic receptor agonist isoproterenol or the NK1-R agonist substance P (SP) were added to COS-7 cells transiently transfected with either β_2-AR/Rluc (Class A 7TMR) and GFP^2/β-arr2 or NK1-R/Rluc (Class B 7TMR) and GFP^2/β-arr2. The substrate DeepBlueC/Coelenterazine 400A (final concentration 5 μM) was then injected by injector and readings were collected 2 seconds after injections. The signals detected at 395 nm and 515 nm were measured sequentially, and the 515/395 ratios calculated and expressed as a miliBRET level (mBU; BRET ratio x 1,000). Results are expressed as the fold difference between the NK1-R/β-arr2 agonist-induced BRET^2 signal and β_2-AR/β-arr2 BRET^2 value. Data represent the mean $\pm$ S.E.M. of triplicate observations from three to five independent experiments. Adopted from Vrecl et al. [20].

As shown in Figure 3, conserved residues within the *carboxyl-terminus* of the β-arrestins are essential for the interaction with the heavy chain of clathrin [34, 35] and the β_2-adaptin subunit of the AP-2 [11, 36, 37]. The clathrin binding domain of β-arrestin2 was located to a small stretch of residues in the *carboxyl-terminus* of the molecule [35]. Site-directed mutagenesis demonstrated the importance of both hydrophobic residues leucine (L-373), isoleucine (I-374), and phenylalanine (F-376)) and acidic glutamic acid (E-375 and E-377) residues in the β-arrestin molecule for arrestin/clathrin interaction [35]. Two *carboxyl-terminal* arginine residues (R-394 and R-396) were also identified in the β-arrestin2 that mediate β-arrestin/AP-2 interaction [36]. The corresponding residues in β-arrestin1 and human β-arrestin2 isoform 1 are R-393 and R-395. Subsequent mutational analysis showed that conserved R-395 is important for interaction with the β_2-adaptin subunit of the AP-2 complex, while the R-393 (homolog of R-382 in visual arrestin) is an integral part of the polar core and thus it is unlikely to mediate binding to other proteins in a basal state [38]. Considering mutagenesis data, β-arr2 mutants deficient in their ability to interact with either the AP-2 (β-arr2 R393E,R395E, β-arr2 394 stop and β-arr2 383 stop) or AP-2 and clathrin (β-arr2 373 stop) were generated (see Figure 3).

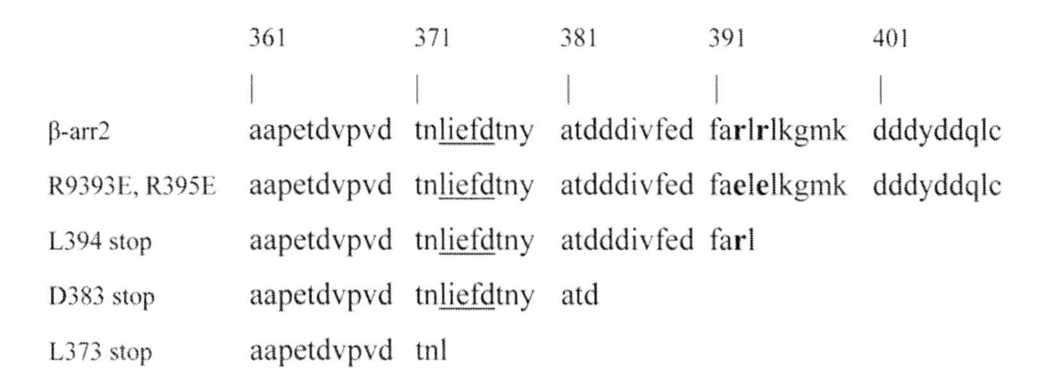

Figure 3. Structure of the carboxyl-terminal domains of the of the human β-arrestin2 isoform 1 and β-arrestin2 mutants. The LØXØ(D/E) motif representing the clathrin binding site in β-arrestin2 (residues 373–377) is underlined and arginine residues (R-393 and R-395) important for the interaction with the β2 subunit of the adaptor protein 2 (AP-2) are in boldface type. Residues in β-arr2 are numbered from amino acids 361 to 409. Ø is a bulky hydrophobic residue and X represents any polar amino acid.

BRET^2 Signal - Effect of Beta-Arrestin2 Mutants

The effectiveness of generated β-arrestin2 fusion constructs in BRET^2 assay were tested using β_2-AR and NK1-R as a model Class A and B 7TMRs, respectively [20]. As depicted in Figure 4, tested β-arr2 mutants had little or no significant effect on the agonist-promoted BRET^2 signal obtained with the NK1-R. This result is as expected, since this receptor class form stable complexes with WT β-arrestins and the resulting high BRET^2 signal could therefore not be further enhanced. However, agonist-promoted interaction of the Class A β_2-AR with either β-arr2 383 stop, β-arr2 373 stop or β-arr2 R393E,R395E resulted in an almost 2.5-fold increase in the BRET^2 signal when compared to either WT β-arr2 or β-arr2 394 stop mutant (Fig. 4). A similar increase in the BRET^2 signal was reported for the neuropeptide Y receptor type 2 (NPY2-R) and the eicosanoid receptor denoted as TG1019 [20]. The enhanced BRET^2 signal observed for the β-arr2 383 stop, β-arr2 373 stop or β-arr2 R393E,R395E mutants was most likely caused by preventing the interaction of the receptor/β-arr2 complex with components of the clathrin coated pit and consequently prolonged the lifetime of the Class A receptor/β-arr2 complex. Disruption of the β-arr2/AP-2 binding (R395E) and reversing the charge of an amino acid in the polar core (R393E; [38]) or removing the AP-2 binding domain (β-arr2 383 stop) resulted in a substantially increased BRET^2 ratio. Additional removal of the clathrin-binding domain did not result in a cumulative effect on the

$BRET^2$ ratio, in accordance with the suggestion that AP-2 interaction precedes binding to clathrin [37]. All tested β-arrestin2 mutants with the exception of β-arr2 394 stop, are capable of tight phosphorylation-independent binding to the receptor since they miss a part or the entire regulatory arrestin carboxyl terminus, which keeps arrestin in a basal conformation [39]. The inability of phosphorylation-dependent β-arr2 394 stop mutant to enhance $BRET^2$ signal could suggest that phosphorylation is a limiting step for β-arr2 binding to β_2-AR in COS-7 cells, but not in HEK-293 (unpublished observation) with relatively high endogenous expression of G protein-coupled receptor kinases (GRKs) [40]. This is also in agreement with previous report stating that only the R-395 and not R-393 is important for β-arrestin interaction with the β_2-adaptin subunit of the AP-2 [38].

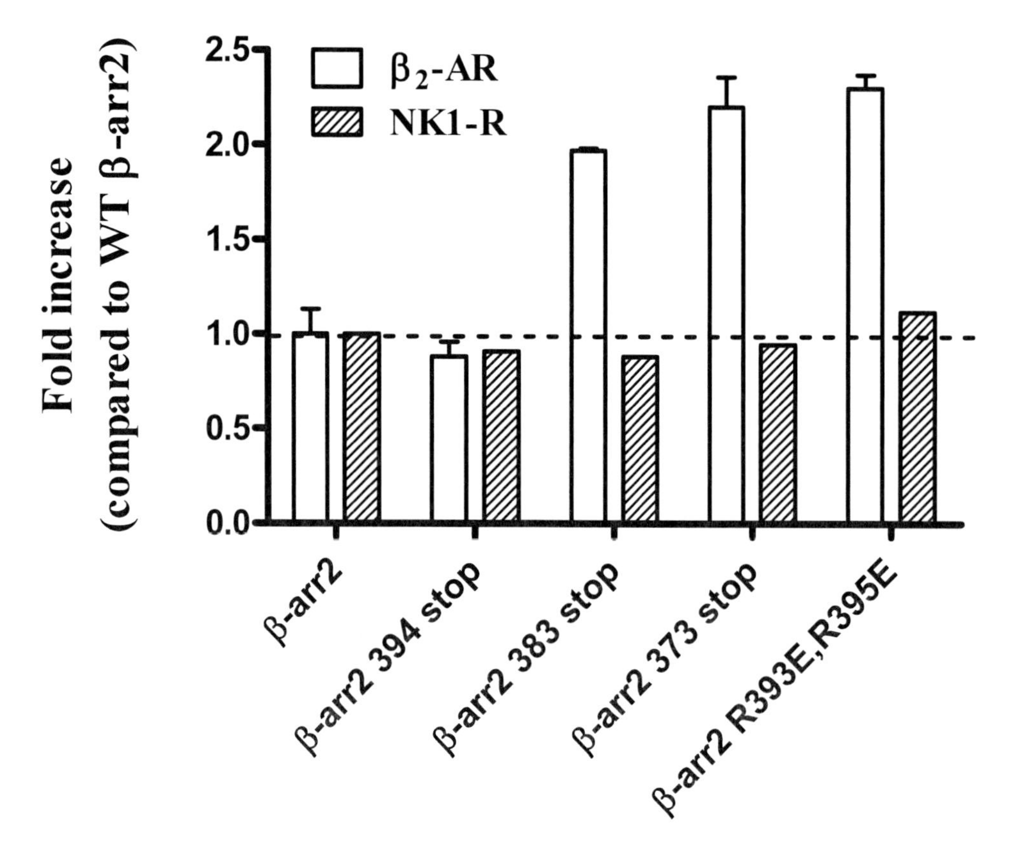

Figure 4. Effectiveness of β-arr2 mutants in comparison to wt β-arr2 in inducing $BRET^2$ signal in COS-7 cells generated by the agonist-induced interaction with Class A and Class B 7TMRs. COS-7 cells were transiently transfected with indicated GFP^2/β-arr2 construct together with either β_2-AR/Rluc (Class A 7TMR) or NK1-R/Rluc (Class B 7TMR). Following harvesting, ~200,000 cells were distributed to wells in 96-well microplates and were treated with either the adrenergic receptor agonist isoproterenol or the NK1-R agonist substance P (SP) and incubated for 5 minutes at room temperature. The substrate DeepBlueC/Coelenterazine 400A (final concentration 5 μM) was then injected by injector and readings were collected 2 seconds after injections. The signals detected at 395 nm and 515 nm were measured sequentially, and the 515/395 ratios calculated and expressed as a miliBRET level (mBU; BRET ratio x 1,000). Results are expressed as the fold increase of mutant β-arr2/receptor $BRET^2$ signal compared to WT β-arr2/receptor $BRET^2$ value. Data represent the mean $\pm$ S.E.M. of triplicate observations from three to five independent experiments.

Combinations of a BRET reader equipped with on-line injectors and selected β-arrestin2 mutants such as β-arr2 R393E,R395E made it possible to perform full-plate screening assays. The use of on-line injectors is essential because the half-life ($t_{1/2}$) of the reaction between DeepBlueC™ and Rluc is very short (e.g. for the assay described here the optimal reading time after DeepBlueC™ injection is 2 sec). The $BRET^2$ signal obtained for the NK1-R and β_2-AR screening assays using the β-arr2 R393E,R395E mutant resulted in robust signals with Z' values in the vicinity of 0.6, suggesting that modified $BRET^2$ assay should be an attractive alternative to existing functional cell-based screening assays, offering a robust and universally applicable assay [6, 20]. For the β_2-AR, the $BRET^2$ assay was also shown superior for secondary screening of agonists where a separation of full and partial agonists is needed, while functional cell-based ALPHAscreen™ cAMP assay may be preferred for primary screening of agonists [41].

$BRET^2$ Signal - Effect of *Carboxyl-Tail* Swapping

The *above-described* screening assay gave excellent performance with numerous 7TMRs, however some receptors still interacted weakly with β-arr2 R393E,R395E mutant, resulting in unsatisfactory assay performance. Human cannabinoid receptor 1 (CB1R), which is an important target in the field of obesity, was such an example. Initial experiments revealed that the agonist-induced interaction between β-arr2 and the CB1R resulted in an almost negligible $BRET^2$ signal. Only a slight increase was observed when the β-arr2 R393E,R395E mutant was used [22]. This was intriguing as based on the reported CB1R/β-arr1 binding [42] a Class B 7TMR behavior was anticipated. Mutations in β-arr2 alone are therefore not always sufficient to enhance receptor/arrestin interaction. This could indicate that the CB1R interacts weakly or transiently with β-arr2, or that the distance and/or orientation between RLuc and GFP^2 is not optimal for the generation of a satisfactory $BRET^2$ signal. With certain 7TMRs, β-arrestin dependence was gained by adding/swapping of the carboxyl-termini of the receptors [43-45] and additionally stable receptor/β-arrestins complexes have been achieved by switching the carboxyl-terminal tails of either Class A and class B receptors [12, 46] or between class B receptors [21]. Using this approach, the carboxyl-terminal tail of the CB1R was replaced with the carboxyl-terminal tail of the vasopressin type 2 receptor (V2R) or the bombesin receptor subtype 3 (BRS3). The V2R and BRS3 *carboxyl*-terminal tails contain serine/threonine clusters in appropriate positions (30–40 amino acids relative to the cytoplasmic face of the membrane) that underlie the stability of the formed receptor/arrestin complex [46, 47] and additionally, the BRS3 receptor was the most potent receptor tested in the running $BRET^2$-based screening platform.

Figure 5 shows sequence alignments of the carboxyl-terminal domains of the CB1R, V2R and BRS3 and generated tail-swapped CB1R chimeric constructs (CB1R-V2R tail and CB1R-BRS3 tail). Note the difference in the positions of serine/threonine clusters within the carboxyl-terminal domains of the studied receptors (V2R and BRS3 *vs.* CB1R). Serine/threonine clusters that mediate the formation of stable receptor/β-arr complexes are present within the CB1R carboxyl-terminal tail, however, based on the previous findings [47, 48], their position is not appropriate to support this role.

A

CB1R NPIIYALRSKDLRHAFRSMFPSĊEGTAQPLDNSMGDSDCLHKHANNAASVHRAAESCIKS**TVK**IAKVTMSVSTDTSAEAL

V2R NPWIYASFSSSVSSELRSLLĊĊARGRTPPSLGPQDESCTTASSSLAKDTSS

BRS3 NPFALYWLSKSFQKHFKAQLFĊĊKAERPEPPVADTSLTTLAVMGTVPGTGSIQMSEISVTSFTGC**SVK**QAEDRF

B

CB1R-V2R tail NPIIYALRSKDLRHAFRSMFPSĊEG-GS-GRTPPSLGPQDESCTTASSSLAKDTSS

CB1R-BRS3 tail NPIIYALRSKDLRHAFRSMFPSĊEG-GS-ERPEPPVADTSLTTLAVMGTVPGTGSIQMSEISVTSFTGC**SVK**QAEDRF

Figure 5. Structure of the carboxyl-terminal domains of the CB1R, V2R and BRS3 (A) and tail-swapped chimeric CB1R constructs (B). A - The amino acid composition of the human CB1R, V2R and BRS3 carboxyl-terminal tails beginning with the conserved NP(X)2,3Y motif that marks the end of the seventh transmembrane domain. The CKII motifs ((S/T)XX(D/E)) are shadowed, the PKC motifs ((S/T)X(R/K)) are in boldface type and serine/threonine clusters (serine and threonine residues occupying three out of three, three out of four, or four out of five consecutive positions) are underlined. Only one serine/threonine cluster that was identified previously by Oakley et al. [46] to mediate the formation of stable receptor-β-arrestin complexes is underlined in the V2R. Putative sites for palmitoylation are indicated with a point. B - The amino acid composition of the carboxyl-terminal domains of tail-swapped chimeric CB1R constructs. The carboxyl-terminal tail of the CB1R was replaced with the tail of the V2R (G345-Stop372) or the BRS3 (E351-Stop400). Tails were exchanged after the putative amphipathic helix 8 of the CB1R in position T418. A two amino acid (GS) linker between the CB1R and either the V2R or BRS3 carboxyl-terminal tail originates from a BamHI-site (GGA TCC) that was designed to enable the tail swapping. (From Vrecl et al., Journal of Biomolecular Screening (vol. 14, No. 4, pp 371-80). Copyright © 2009 by (SAGE Publications). Reprinted by Permission of SAGE Publications).

The generated receptor chimeric constructs were tested in the BRET^2 assay and receptor interactions with both WT β-arr2 and the β-arr2 R393E,R395E mutant were investigated and compared with the corresponding WT receptors (Figure 6).

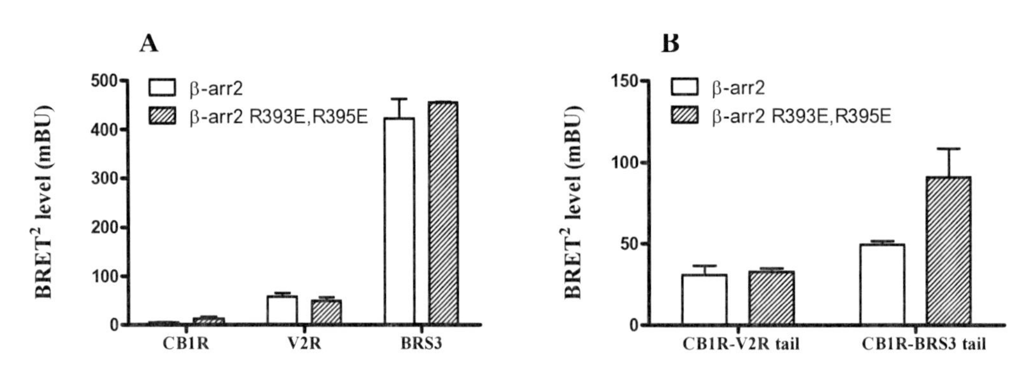

Figure 6. Comparison of agonist-induced BRET^2 signals from WT CB1R, V2R and BRS3 and CB1R tail-swapped chimeric constructs. HEK-293 cells were transiently transfected with GFP^2/β-arr2 (open bars) or GFP^2/β-arr2 R393E,R395E (hatched bars) together with the indicated (A) WT or (B) CB1R tail-swapped chimeric RLuc-tagged receptor constructs. Following harvesting, ~200,000 cells were distributed to wells in 96-well microplates and were treated with appropriate agonists (CP 55,940, [Arg8]-VP or BRS3 Analog 1; 10^{-6} M final concentration) and incubated for 5 min at room temperature. The substrate DeepBlueC/Coelenterazine 400A (final concentration 5 μM) was then injected by injector and readings were collected 2 sec after injections. The signals detected at 395 nm and 515 nm were measured sequentially, and the 515/395 ratios calculated and expressed as a miliBRET level (mBU; BRET ratio x 1,000). Data represent the mean ± S.E.M. of triplicate observations from three to five independent experiments.

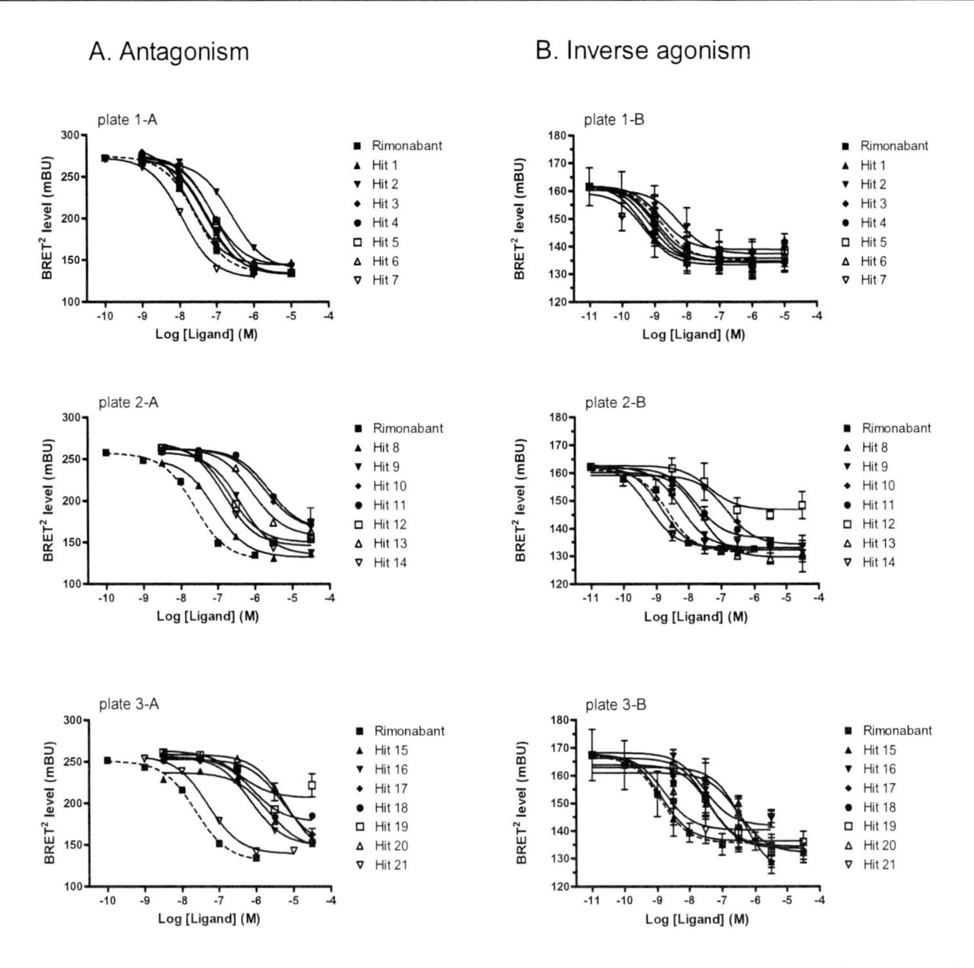

Figure 7. BRET2-based secondary antagonist and inverse agonist dose response curves for selected hits. A - Secondary antagonist screening (antagonist dose-response curves) was performed for 21 compounds identified in primary antagonist screening. First, 1 µl of increasing concentrations of the test compound ranging from 10−10 to 5x10−5 M (final concentration) and 1 µl of CB1R agonist CP 55,940 (50 nM final concentration), both dissolved in DMSO, were transferred into wells of 96-well microplates (white Optiplate) using a Biomek FX. Negative and positive controls were in columns 1 (wells 1-8) and 12 (wells 89 -96). Two wells in each column contained 1 µM CP 55,940, three wells contained 50 nM CP 55,940 (EC80 concentration) and three wells contained DMSO (1 vol %). Furthermore, a rimonabant dose-response curve (each concentration in duplicate) (10−10 to 10−6 M) was prepared in row one. B - The selected compounds were also tested in inverse agonism mode. First, 1 µl of increasing concentrations of the test compounds ranging from 10−10 to 5x10−5 M (final concentration) dissolved in DMSO was transferred into wells of 96-well microplates (white Optiplate) using a Biomek FX. Negative (1 vol % DMSO) and positive (1 □M rimonabant) controls were in columns 1 (wells 1-8) and 12 (wells 89 -96). Furthermore, a rimonabant dose-response curve (each concentration in duplicate) (10−10 to 10−6 M) was prepared in row one. Subsequently, plates were transferred to the Mithras LB 940 reader, where 180 □l of re-suspended cells containing ~200,000 cells were distributed in 96-well microplates by the Mithras injector 1. Substrate Coelenterazine 400a (final concentration 5 □M) was then injected by injector 2 and readings were collected 2 sec after injections. The signals detected at 395 nm and 515 nm were measured sequentially, and the 515/395 ratios calculated and expressed as a mBU (BRET ratio x 1,000). Data represent the mean + S.E.M. of duplicate observations from a single representative screening experiment. (From Vrecl et al., Journal of Biomolecular Screening (vol. 14, No. 4, pp 371-80). Copyright © 2009 by (SAGE Publications). Reprinted by Permission of SAGE Publications).

The $BRET^2$ signal for the CB1R was greatly improved with the replacement of its carboxyl-terminal tail with that of the V2R or BRS3. It was also observed that co-expression with the β-arr2 R393E,R395E mutant considerably improved the $BRET^2$ signal in the case of the CB1R-BRS3 tail construct, while the same was observed neither for the V2R nor for the CB1R-V2R tail chimera (compare Figure 6A and B).

It is likely that the relative position of the serine/threonine clusters that are 39 and 31 amino acids downstream of the NPXXY motif for the CB1R-V2R tail and CB1R-BRS3 tail chimera, respectively, caused a difference in the efficacy of the chimeric receptors. Based on these data, the CB1R-BRS3 tail construct was selected in combination with the HEK-293 stable cell line that expresses a high and constant level of the GFP^2/β-arr2 R393E,R395E mutant for subsequent screening assays.

Due to the hydrophobic nature of the CB1R ligands and technical limitations connected with the existing instrumentation (i.e. minimal volume injected by injectors), the following modifications were necessary to run CB1R screening assay in fully automated mode. It was demonstrated that pre-treatment with antagonist/test compound is not necessary to run primary or secondary antagonist screening, but that the cells can be incubated concomitantly with agonist and antagonist/test compounds. A library of 260 compounds was screened and 21 antagonist and inverse agonist hits identified with IC_{50} and EC_{50} values ranging from 0.3 nM to 7.5 μM. Both primary and secondary screening were performed with $Z' > 0.5$ [22]. Examples of secondary antagonist screening ($BRET^2$ antagonist dose-response curves) and inverse agonist dose response curves are shown in Figure 7.

The assay window is smaller in inverse agonism screening compared to the secondary screening (≈30 *vs.* ≈100 mBU), which is consistent with the constitutive and maximal agonist-induced $BRET^2$ signal. Calculated IC_{50} and EC_{50} values revealed that tested hits displayed higher potency when tested in inverse agonism mode than in antagonist mode; however, the rank order of potency remained comparable. The rank order of potencies obtained in $BRET^2$ assays also correlates well with that obtained in the classical [^{35}S]-GTPγ-S assay. Thereby, this assay allows rapid discrimination between neutral antagonists and inverse agonists [22].

Other existing cell-based assays used for screening the CB1R are GTPγ-S binding *and IP assays using promiscuous Gα proteins [7]. However, the agonist-induced assay window is relatively small in both assays due to the constitutive activity of the CB1R and relatively low amplification in the signaling cascade (e.g. in striatum each CB1R activates only 3 G-proteins) (reviewed in [49]).* The applicability of the tail swap strategy was also proven successful with the chemokine receptor type 4 [21] and melanocortin-4 (MC4) receptor (7TM Pharma; unpublished).

Conclusion

The growing number of drug targets and the complexity of the intracellular events following ligand/receptor interaction represent a pressing demand and a technological challenge for the development of a generally applicable, functional, cell-based screening platform. Recent advances in detection technologies and technological innovation were instrumental for turning the BRET method into a new technological platform suitable for both research and compound MHT/HTS screening. BRET^2 receptor/β-arrestin screening assay decribed here is a universal, functional screening assay that is reliable, powerful, time- and resource-efficient as well as user- and environmental-friendly. It is expected that this versatile and highly sensitive screening assay would play an important role in discovering the next generation of pharmaceuticals.

ACKNOWLEDGMENT

The work presented was performed at the 7TM Pharma A/S, Denmark in the frame of Slovenian-Danish collaboration grants D-3/2002, D-04/2005, BI-DK/06-07-007 and BI-DK/07-09-002 supported by the Slovenian Research Agency.

REFERENCES

[1] Lagerstrom MC, Schioth HB. *Nat Rev Drug Discov* 2008;7(4):339-357.
[2] Overington JP, Al-Lazikani B, Hopkins AL. *Nat Rev Drug Discov* 2006;5(12):993-996.
[3] Kenakin TP. *Nat Rev Drug Discov* 2009;8(8):617-626.
[4] Walters WP, Namchuk M. *Nat Rev Drug Discov* 2003;2(4):259-266.
[5] Siehler S. *Biotechnol J* 2008;3(4):471-483.
[6] Heding A. *Expert Rev Mol Diagn* 2004;4(3):403-411.
[7] Kostenis E, Waelbroeck M, Milligan G. *Trends Pharmacol Sci* 2005;26(11):595-602.
[8] Hebert TE, Gales C, Rebois RV. *Cell Biochem Biophys* 2006;45(1):85-109.
[9] Gurevich VV, Gurevich EV. *Pharmacol Ther* 2006;110(3):465-502.
[10] Goodman OB, Jr., Krupnick JG, Gurevich VV, Benovic JL, Keen JH. *J Biol Chem* 1997;272(23):15017-15022.
[11] Laporte SA, Oakley RH, Zhang J, Holt JA, Ferguson SS, Caron MG, Barak LS. *Proc Natl Acad Sci U S A* 1999;96(7):3712-3717.
[12] Oakley RH, Laporte SA, Holt JA, Caron MG, Barak LS. *J Biol Chem* 2000;275(22):17201-17210.
[13] Dale LB, Bhattacharya M, Seachrist JL, Anborgh PH, Ferguson SS. *Mol Pharmacol* 2001;60(6):1243-1253.
[14] Tulipano G, Stumm R, Pfeiffer M, Kreienkamp HJ, Hollt V, Schulz S. *J Biol Chem* 2004;279(20):21374-21382.
[15] Fairfax BP, Pitcher JA, Scott MG, Calver AR, Pangalos MN, Moss SJ, Couve A. J Biol Chem 2004;279(13):12565-12573.
[16] Vrecl M, Anderson L, Hanyaloglu A, McGregor AM, Groarke AD, Milligan G, Taylor PL, Eidne KA. *Mol Endocrinol* 1998;12(12):1818-1829.
[17] Oakley RH, Hudson CC, Cruickshank RD, Meyers DM, Payne RE, Rhem SM, Loomis CR. *Assay Drug Dev Techn* 2002;1:21-30.
[18] [Ferguson SS, Caron MG. *Methods Mol Biol* 2004;237:121-126.
[19] Bertrand L, Parent S, Caron M, Legault M, Joly E, Angers S, Bouvier M, Brown M, Houle B, Menard L. *J Recept Signal Transduct Res* 2002;22(1-4):533-541.
[20] Vrecl M, Jorgensen R, Pogačnik A, Heding A. *J Biomol Screen* 2004;9(4):322-333.

[21] Hamdan FF, Audet M, Garneau P, Pelletier J, Bouvier M. *J Biomol Screen* 2005;10(5):463-475.
[22] Vrecl M, Norregaard PK, Almholt DL, Elster L, Pogačnik A, Heding A. *J Biomol Screen* 2009;14(4):371-380.
[23] Olson KR, Eglen RM. *Assay Drug Dev Technol* 2007;5(1):137-144.
[24] Zhao X, Jones A, Olson KR, Peng K, Wehrman T, Park A, Mallari R, Nebalasca D, Young SW, Xiao SH. *J Biomol Screen* 2008;13(8):737-747.
[25] Barnea G, Strapps W, Herrada G, Berman Y, Ong J, Kloss B, Axel R, Lee KJ. *Proc Natl Acad Sci U S A* 2008;105(1):64-69.
[26] Zacharias DA, Baird GS, Tsien RY. *Curr Opin Neurobiol* 2000;10(3):416-421.
[27] Xu Y, Piston DW, Johnson CH. *Proc Natl Acad Sci U S A* 1999;96(1):151-156.
[28] Angers S, Salahpour A, Joly E, Hilairet S, Chelsky D, Dennis M, Bouvier M. *Proc Natl Acad Sci U S A* 2000;97(7):3684-3689.
[29] Xu Y, Kanauchi A, von Arnim AG, Piston DW, Johnson CH. *Methods Enzymol* 2003;360:289-301.
[30] De A, Loening AM, Gambhir SS. *Cancer Res* 2007;67(15):7175-7183.
[31] De A, Ray P, Loening AM, Gambhir SS. *Faseb J* 2009;23(8):2702-2709.
[32] Pfleger KD, Dromey JR, Dalrymple MB, Lim EM, Thomas WG, Eidne KA. *Cell Signal* 2006;18(10):1664-1670.
[33] Boute N, Pernet K, Issad T. *Mol Pharmacol* 2001;60(4):640-645.
[34] Goodman OB, Jr., Krupnick JG, Santini F, Gurevich VV, Penn RB, Gagnon AW, Keen JH, Benovic JL. *Nature* 1996;383(6599):447-450.
[35] Krupnick JG, Goodman OB, Jr., Keen JH, Benovic JL. *J Biol Chem* 1997;272(23):15011-15016.
[36] Laporte SA, Oakley RH, Holt JA, Barak LS, Caron MG. *J Biol Chem* 2000;275(30):23120-23126.
[37] Laporte SA, Miller WE, Kim KM, Caron MG. *J Biol Chem* 2002;277(11):9247-9254.
[38] Milano SK, Pace HC, Kim YM, Brenner C, Benovic JL. *Biochemistry* 2002;41(10):3321-3328.
[39] Kovoor A, Celver J, Abdryashitov RI, Chavkin C, Gurevich VV. *J Biol Chem* 1999;274(11):6831-6834.
[40] [Menard L, Ferguson SS, Zhang J, Lin FT, Lefkowitz RJ, Caron MG, Barak LS. *Mol Pharmacol* 1997;51(5):800-808.
[41] Elster L, Elling C, Heding A. *J Biomol Screen* 2007;12(1):41-49.
[42] Bakshi K, Mercier RW, Pavlopoulos S. *FEBS Lett* 2007;581(25):5009-5016.
[43] Heding A, Vrecl M, Bogerd J, McGregor A, Sellar R, Taylor PL, Eidne KA. *J Biol Chem* 1998;273(19):11472-11477.
[44] [Heding A, Vrecl M, Hanyaloglu AC, Sellar R, Taylor PL, Eidne KA. *Endocrinology* 2000;141(1):299-306.
[45] Hanyaloglu AC, Vrecl M, Kroeger KM, Miles LE, Qian H, Thomas WG, Eidne KA. *J Biol Chem* 2001;276(21):18066-18074.
[46] Oakley RH, Laporte SA, Holt JA, Barak LS, Caron MG. *J Biol Chem* 1999;274(45):32248-32257.
[47] Oakley RH, Laporte SA, Holt JA, Barak LS, Caron MG. *J Biol Chem* 2001;276(22):19452-19460.

[48] Qian H, Pipolo L, Thomas WG. *Mol Endocrinol* 2001;15(10):1706-1719.
[49] Howlett AC, Breivogel CS, Childers SR, Deadwyler SA, Hampson RE, Porrino LJ. *Neuropharmacology* 2004;47 Suppl 1:345-358.

In: Bioluminescence
Editor: David J. Rodgerson, pp. 153-166
ISBN 978-1-61209-747-3

Chapter 9

Characteristics and Control of Bacterial Bioluminescence

Satoshi Sasaki
School of Bioscience and Biotechnology,
Tokyo University of Technology,
Hachioji, Japan

Abstract

As luminescent bacteria can convert chemical energy into light using luciferase, they are known to glow in the dark with a visible peak wavelength (in the case of *Photobacterium kishitanii*, ca. 475 nm). The organism is thought to have luciferase, evolutionally, to scavenge oxygen. Industrially, luminescent bacteria is used for toxicity measurement. In this system, luminescence intensity is inhibited by the toxic compounds in the sample. Stabilization of the luminescence intensity is therefore important to realize highly sensitive measurement. Experimentally, control of luminescence intensity is not easy. For example, oscillation in the luminescence intensity from the bacterial suspension is often observed in a certain environment. Reaction-diffusion of dissolved oxygen into the cell, and synchronization of luciferase gene expression as a result of quorum sensing, were thought to be two reasons for the oscillation. To explain such a behavior, the characteristics of the cells in suspension was investigated using a microfluidic device. In order to compare the luminescent intensity of bioluminescence from marine luminous bacteria with different motility, luminescent bacteria were separated according to their motility using the device. Calculation of the luminescent intensity per cell was performed, and swimmer cells were shown to be brighter that the others. Luminescence from intact bacteria also show interesting characteristics such as color changing or irradiation-controlled quenching. Understanding of such characteristics will be a key for a novel application. In this chapter, recent findings in the bacterial luminescence is reported, and their application is proposed.

1. Introduction

Organisms that glow in the night have been known for thousands of years from Greek philosopher Aristotle's report. Luminescent insect such as firefly often appears in the ancient poetry. One of 54 chapters of the ancient novel, The Tale of Genji, written one thousand years ago, was entitled exactly the same name with the insect. In the literature, bioluminescence is often regarded as something fragile, strongly in connection with the transience of life. It was believed that after passing away the human's spirit escapes from the body and flies around in the air glowing. Apart from such unidentified phenomena, glowing of uncooked seafood or luminescence of large sea area are now well understood as the result of the growth of bioluminescent bacteria in large density.

Biologically, bioluminescent organisms are found in several kingdoms. Biolumionescent bacteria are the typical example in the Kingdom Monera. In the Kingdom Fungi, luminous fungi such as *Mycena chlorophos* is well known. *Noctiluca scintillans*, known as sea sparkle, is an luminescent organism belong to the Kingdom Protista. Firefly is one of the well known insect that belongs to the Kingom Animalia. So far, on the other hand, no organism is reported in the Kingdom Plantae that show bioluminescence. In general, we find variety of bioluminescent organism in the sea more than in fresh water [1].

2. Bioluminescent Bacteria

Bioluminescent bacteria are well-known marine organism for emitting blue-green light. The bacteria have historically been called in various names. A free-living species *Vibrio harveyi* and bioluminescent species in *Shewanella* have been well known, but they have been sometimes called *Photobacterium phosphoreum*. *Vibrio fisheri* and *Photobacterium leiognathi* are involved in symbiosis, former with squid and the latter with fish. Fish or squid for sashimi (uncooked) glow when soaked overnight in saline due to this symbiosis. Bacterial luciferase is responsible for the bioluminescent reaction. In the reaction, oxidation of $FMNH_2$ (eqn. 2.1) and a long-chain fatty aldehyde occurs at the same time. The mechanism of such bioluminescence has been applied to biotechnological fields [2,3]. From the viewpoint of chemical energy consumption, bacterial bioluminescence can be referred to as an extravagant system for light production, where 200 kJ of energy is consumed to emit 1 Einstein of photon [4]. This amount of chemical energy might be stored at the beginning of respiratory chain in the form of NADH as a reducing potential. Bacterial bioluminescent is therefore closely related to the respiration of the organism. The reaction involved in the bioluminescence can be written as

$$FMNH_2 + O_2 + RCHO \xrightarrow{luciferase} FMN + H_2O + RCOOH + light \qquad (2.1)$$

where RCHO and RCOOH represents an aldehyde and a carboxylic acid, respectively. Reduced type flavin mononucleotide ($FMNH_2$) appears in the reaction as a substrate together with O_2 and RCHO. The rate of reaction that is strongly related to the light intensity therefore depends on the concentrations of the three substrates. Kinetically, under constant luciferase activity, the rate-determining factor will be the substrate with minimal concentration: if one of the substrate is consumed too much, then the light intensity depends on the substrate concentration. Flavin is reduced inside the bacterial cells through the catalysis by enzymes in the NAD(P)H−flavin oxidoreductase (flavin reductase) family at the expense of NAD(P)H [5]. NADH is also oxidized in the bacterial respiratory chain to obtain energy to synthesize ATP. Generally, the oxidization reaction is exoergonic, where ca. 220 kJ mol^{-1} of energy is produced. This energy is consumed for the synthesis of several moles of ATP. Here we can see that energy produced through the oxidation of NADH is shared by two process; ATP synthesis and bioluminescence. Bioluminescence is therefore extravagant. Bioluminescent bacteria emit light with all their might! 20% of oxygen adsorbed by the microbe is consumed to produce light [6], and 5% of the soluble protein in the cell is luciferase. In fact, bioluminescence intensity was reported to increase when the respiratory chain is suppressed by the addition of KCN [7].

The amount of luciferase in a cell is not constant; the enzyme is synthesized through the expression of series of genes. In the case of *Vibrio fischeri*, they are called the lux gene. They are composed of *lux*AB, *lux*CDE and *lux*G that is involved in the synthesis of luciferase αβ subunit, aldehyde synthase, and FMN reductase, respectively. This set of genes are called *lux*CDABEG operon. Two more genes, *lux*R and *lux*I compose the lux gene together with the operon. *lux*R is involved in the synthesis of the protein luxR that works as an activator for the transcription of the operon. *lux*I, on the other hand, is concerned with the synthesis of autoinducer, which serves as a cell-cell communication molecule.

In the case of *V. fischeri*, this transcriptional coregulator of the lux regulon has been identified as *N*-(3-oxohexanoyl) homoserine lactone [8]. When the lacton molecule bind to the luxR protein, the Combination of the two (AI-luxR) binds the upstream of *lux*CDABEG operon, resulting in the luciferase synthesis. Thus, bacterial luminescence ability depends on the cell density of the bacterial suspension. At low cell density luciferase is not synthesized. Only after the cell density exceeded a certain threshold as a result of cell division, the bacteria start to synthesize luciferase and its substrates, resulting in luminescence. Such a regulation of gene expression in response to fluctuations in cell-population density is called quorum sensing. Accordingly, like many other bacteria [9], *V. fischeri* produce and release N-(3-oxohexanoyl) homoserine lactone that increase in concentration as the cell grows. The detection of a minimal threshold concentration of this autoinducer molecule leads to the expression of *lux*CDABEG that is involved in luciferase and substrate synthesis [10].

Here we assume that bioluminescence occurs only through the catalysis of luciferase. It is therefore possible, if we assume luciferaes amount to be constant, to quantify the luminescence intensity using a simple model as following. Generally, in the simplest Uni Uni enzymatic reaction system where a single substrate goes to a single product, the reaction sequence can be written as

$$E+S \underset{k-1}{\overset{k1}{\rightleftarrows}} ES \underset{k-2}{\overset{k2}{\rightleftarrows}} EP \underset{k-3}{\overset{k3}{\rightleftarrows}} E+P \qquad (2.2)$$

where ES and EP represent central complexes. For simplicity let us consider that the reverse reaction from the product P can be neglected, because once the light is emitted it will never come back to start the reverse reaction. Let us also assume that there is one central complex. Then the reaction can be written:

$$E + S \underset{k-1}{\overset{k1}{\rightleftarrows}} ES \xrightarrow{kp} E + P \tag{2.3}$$

What happens when we can assume that E, S and ES equilibrate very rapidly than the rate of ES breakdown into E+P? This is the so-called rapid equilibrium condition, and the rate determining step is the one at which ES breaks down to E+P. Reaction velocity, therefore, can be written by using the catalytic rate constant k_p:

$$v = k_p[ES] \tag{2.4}$$

Total amount of enzyme equals the sum of E and ES, so

$$[E]_t = [E] + [ES] \tag{2.5}$$

Then the equation (2.4) is written:

$$\frac{v}{[E]_t} = \frac{k_p[ES]}{[E]+[ES]} \tag{2.6}$$

By introducing K_S, as defined as

$$K_S = \frac{[E][S]}{[ES]} = \frac{k_{-1}}{k_1} \tag{2.7}$$

we will have

$$\frac{v}{k_p[E]_t} = \frac{[S]}{K_s + [S]} \tag{2.8}$$

This equation can be rewritten as

$$v = \frac{k_p[E]_t[S]}{K_s + [S]} \tag{2.9}$$

These results were obtained according to the well known approach by Henri, Michaelis and Menten. In the case where the E+P forming rate from ES is assumed to be much rapid compared to the rate of ES→E + S, a different approach by Briggs and Haldane is useful. In this case, also, the reaction velocity is written as the similar form such as

$$v = \frac{k_p [E]_t [S]}{K_m + [S]} \tag{2.10}$$

where Km can be written as

$$K_m = \frac{k_{-1} + k_p}{k_1} \tag{2.11}$$

Bioluminescence intensity I_0 is linearly related to the reaction velocity. So, when the amount of enzyme is much larger than that of substrate, the intensity would be written, by using a certain constant K (K=K_s or K_m), as

$$I_0 = \frac{k_p [E]_t [S]}{K + [S]} = \frac{k_p [S]}{\frac{K}{[E]_t} + \frac{[S]}{[E]_t}} \approx \frac{k_p [S]}{\left(\frac{K}{[E]_t} \right)} = \frac{k_p}{K} [E]_t [S] \tag{2.12}$$

What the equation tell us is that at the condition [S] << [E] bioluminescence intensity would be linearly related to the substrate concentration.

If we are to consider bioluminescence in a time scale where cell division effect cannot be neglected, i.e. where [E] and [S] dynamically changes, we have to choose a different approach to estimate I_0 as a function of time.

3. Stability of Bioluminescence Intensity

As a living organism, a bioluminescence bacterium divides into two cells during the growth. Doubling time of the bacteria is in the range of several tens of minutes to hours [11,12]. In several hours, therefore, cell number as well as autoinducer concentration increase and as a result total luciferase amount would change, whether in the form of E or ES. Increase in luciferase amount would be initially controlled by the quorum sensing that is depended on the cell density. As mentioned, bioluminescence is an energy-consuming process that uses $FMNH_2$, O_2 and RCHO as fuels. When a bacterium glow very strong until certain intensity, then it consumes substrate completely and be exhausted.

Consumption of O_2 would be remarkably rapid because this substrate is supplied from outside the cell. The bacterium cannot keep on glowing as it no longer has any O_2 in it. Additionally, the lack of O_2 prevents the bacterium from producing ATP through the respiratory chain. In other words, the bright bacterium is too "tired out" to grow into two cells. Thus, once a bacterium emit light at certain intensity, the cell then cease to glow and stay without divide into two.

Here let us focus on the luminescence behavior of a single glowing bacterium. Let us define the bioluminescence intensity of the bacteria at time t as $I(t)$. In a time period Δt, bacteria number would increase and as a result the total luminescence intensity also would be larger. In the case where the bacterial cells contain same amount of all three substrates even after the cell division, we can expect that the increase in the luminescence intensity, $\Delta I(t)$ is in proportion to the cell number, which is also in proportion to the intensity $I(t)$. Using a coefficient α, $I(t)$ increasing rate can be written as

$$\frac{\Delta I(t)}{\Delta t} = \alpha I(t) \tag{3.1}$$

If we are allowed to extend the time period into a more shorter period dt, we obtain

$$\frac{dI(t)}{dt} = \alpha I(t) \tag{3.2}$$

As mentioned above, after glow the bacterium might lose much of the ability to glow. The α value would then not be constant; it would depend on $I(t)$. Two factors could be considered to see the form of α. Firstly, when the bacteria density exceeded a certain threshold, luciferase is newly synthesized and works for the increase in $I(t)$. This phenomena can roughly be written mathematically that when $I(t)$ exceeded a certain intensity a, quorum sensing occurs and $I(t)$ increases. Secondly, when $I(t)$ exceeds a certain value b, no substrate is left in the cell, resulting in the decrease in $I(t)$ after the time period of dt. The coefficient α can therefore be written as

$$\alpha = \{I(t) - a\}\{b - I(t)\} \tag{3.3}$$

We finally get

$$\frac{dI(t)}{dt} = \{I(t) - a\}\{b - I(t)\}I(t) \tag{3.4}$$

This is just a very rough assumption of the bacterial bioluminescence. It is not justified at all at the moment, but we might have a rough image of the dynamics of the luminescence by solving this differential equation.

4. Analysis of The Stability

What function $I(t)$ look like when it is expressed as the equation (3.4)? When we write $I(t) = x$, the equation becomes

$$\frac{dx}{dt} = mx(b-x)(x-a) \tag{4.1}$$

Steady state, where x do not seem to change with time, is observed when $\frac{dx}{dt}$ equals zero. Such condition is satisfied when $mx(b-x)(x-a)$=0. Solutions for the equation are $x=0$, $x=a$, and $x=b$. When we draw a graph of $y = mx(b-x)(x-a)$ assuming that $0 < a < b$, Fig. 1 is obtained.

To understand the meaning of the equation (4.1), it is sometimes effective to interpret the equation as a vector field [13]. Let us plot $\frac{dx}{dt}$ versus x, and at the same time draw arrows on the x-axis to show if the value $\frac{dx}{dt}$ is positive or negative (if positive, $\rightarrow$. negative, $\leftarrow$).

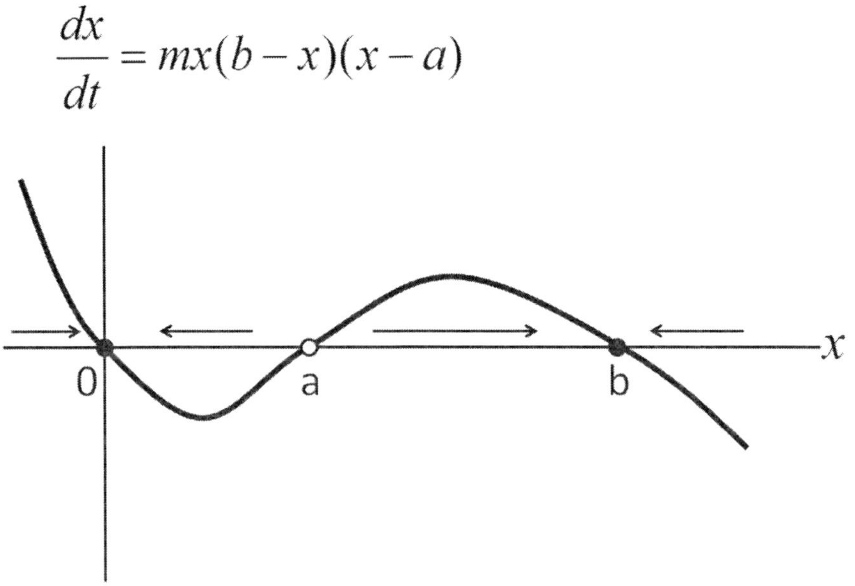

Figure 1. Curve of $\frac{dx}{dt}$ plotted versus x. Closed circles represent stable steady states and open circle an unstable state. Arrow show the directions of tendencies that x would change.

Arrows in the Fig. 1 help us to imagine whether x decrease or increase. Let us consider that the value x changes from 0 to b (let $a < b$). When the x value is smaller than a, the value $\frac{dx}{dt}$ is negative and x will decrease as time. When x, on the other hand, becomes large enough to exceed the value a, then the value $\frac{dx}{dt}$ becomes positive and x is in turn expected to increase as time until x equals to b. Even if x becomes larger than b, $\frac{dx}{dt}$ becomes negative and x goes back to the value b. Through such thinking $x = 0$ and $x = b$ are the stable values, whereas $x = a$ is an unstable one.

We are interested in the fate of x, i.e. the time course of x. Fig. 1 means that if $0 < x < a$ then x will finally reaches 0, and that if $a < x < b$ then x will finally reaches b Then we can illustrate rough curves of x versus t, regardless of quantity, as in Fig. 2.

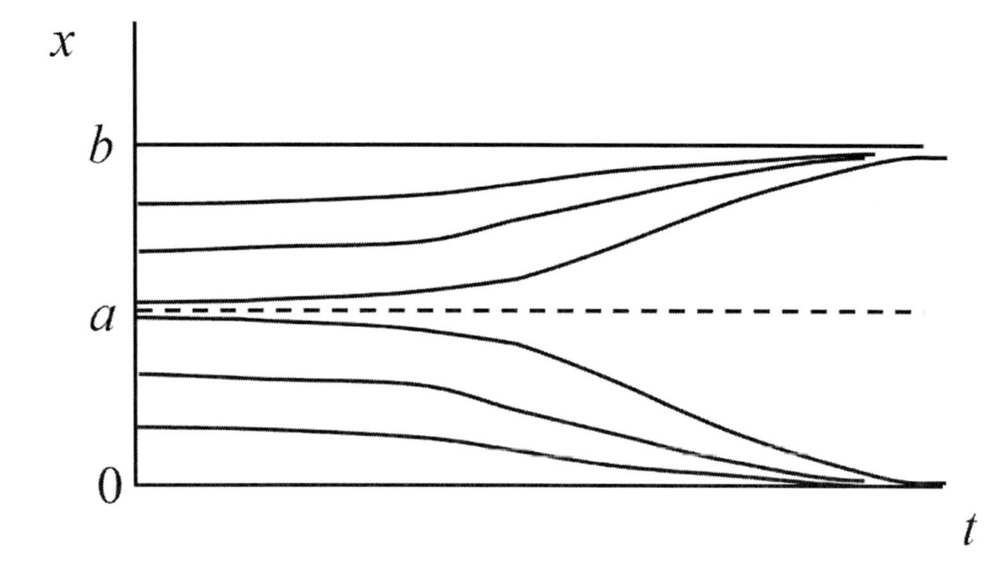

Figure 2. The solution of the equation (4.1) plotted in qualitative forms versus t.

What Fig. 2 tells us is that there are two stable x values. We can translate this in experimental viewpoint such as that if bioluminescent intensity is larger than a it would increase until it reaches b, and that if smaller than a it would decrease until it becomes zero. Thus in the case where x fluctuates around the value a it will finally be forced in time to take value of either a or b. We can also imagine that if the fluctuation of x is large; frequently changes across the value a, the value x oscillates. Although a single bacterium cell would indicate one single luminescence intensity, in the suspension a huge number of cells with varying cell phases are concerned in the luminescence. Even if one or several cells with similar phases perform luminescence that is characterized by Fig. 2, suspension luminescence might be understood as the sum (linear / nonlinear) of the whole phases and be observed as oscillatory in its intensity.

5. Oscillation in Bacterial Luminescence

In our previous work we found an oscillatory behavior in bacterial bioluminescence. *Photobacterium kishitanii*, named in honor of Japanese scientist Teijiro Kishitani who isolated luminous bacteria for the first time from fish light organ, was used for the experiments. Bioluminescent intensity from the bacterial suspension was for the first time thought to be proportional to the cell density at the early stage of growth, because after quorum sensing occurs luciferase is constantly synthesized inside the bacterial cells. We therefore expected that bioluminescence intensity curve would look like, if not be similar to, that of typical growth curve.

In a batch culture without stirring, luminescence from the bacterial suspension was not homogeneous because of convection or precipitation [14,15]. A well-stirred suspension was employed for the observation. Magnetically stirred suspension showed, at a certain broth contents, oscillatory behavior in the luminescence [16]. In this experiment, suspension with nutritiously poorer broth showed more distinct oscillatory waves. Dissolved oxygen (DO) concentration inside the suspension was measured at the same time, showing no significant relation to the luminescence. Do other bacteria show similar characteristics? So far we have tried the yellow bioluminescence bacterium *Vibrio fischeri* Y1. Two different suspensions, containing *P. kishitanii* and *V. fischeri* Y1 respectively, were prepared using the liquid broth with the same content and incubated for several days. The former species is generally known to glow brighter than the latter and as a result the peak intensity from the former suspension was ca. ten times larger than the latter (Figure 3)

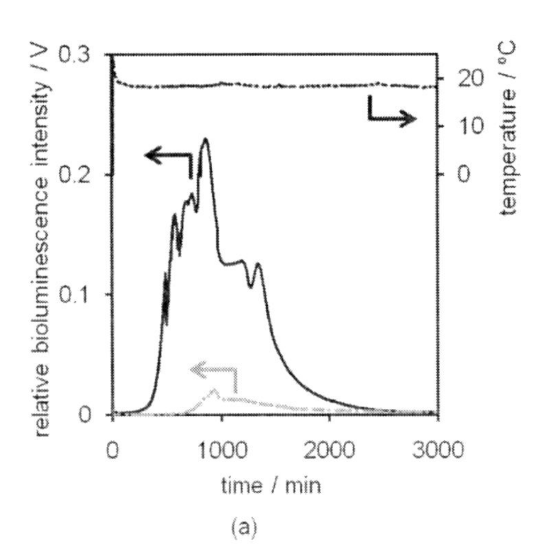

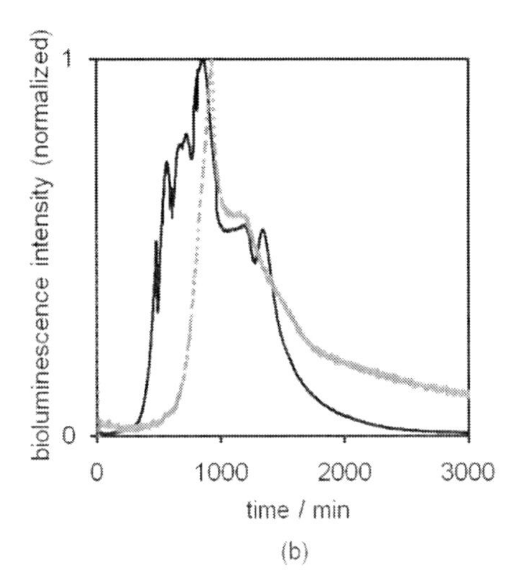

Figure 3. Bioluminescence intensity curves from the suspensions of *P. kishitanii* (solid black line) and of *V. fischeri* Y1 (dotted gray line). Two curves plotted in a same scale (a). Curves plotted in normalized scale (b). Experimentally, 10 mL of suspensions were prepared using Difco™ marine broth 2216, Becton, Dickinson and Company. Both suspensions were well-stirred by magnetic stirrers at 18°C. Intensity was measured using a self-made detector.

The two intensity curves in Fig. 3 (b) shows that bioluminescence from *P. kishitanii* is apparently fluctuated compared to that from *V. fischeri* Y1.

We suspected that bioluminescent intensity depends on DO inside the cell rather than on DO in the suspension, because bacterial suspension was often observed to glow for several hours even when the suspension DO is nearly zero. Pathway of oxygen that is supplied to the bioluminescence reaction inside the bacterial cells should therefore be important to see the oxygen effect on bioluminescence. Luminescence from the bacterial colony on agar plate is reported to depend on oxygen supply [17]. This reversible recovery in the luminescence was understood to be due to the diffusion of oxygen into the bacterial colony. In our work, a similar recovery of luminescence was observed after irradiation (Figure 4). Blue light with peak wavelength of 450 nm was irradiated for several minutes to a single colony of *P. kishitanii* on a agar plate. After the light was turned off, luminescence from the colony decreased, followed by gradual recovery (Figure 4). In a oxygen-free solution, FMN reduction occurs by the long-UV irradiation [18]. In our case, on the other hand, oxygen is thought to be supplied to the colony. Assuming the above mentioned effect of oxygen supply on the colony luminescence, the role of diffused oxygen cannot be neglected during the recovery of luminescence after irradiation.

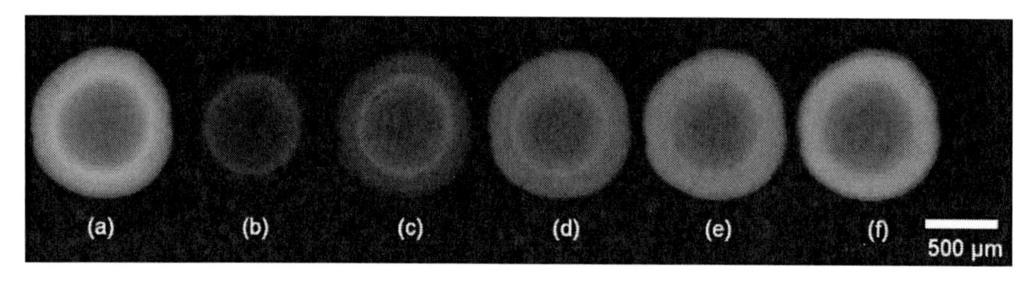

Figure 4. Luminescence images of a single colony of *P. kishitanii* on an agar plate before and after irradiation. Irradiation was performed for 1 min. using the light source of a fluorescence microscope (peak wavelength; 450 nm). Luminescence images before the irradiation (a), 0 min (b), 5 min (c), 10 min (d), 15 min (e) and 20 min (f) after the irradiation were shown.

To see the oxygen effect on the luminescence from different viewpoint, we also investigated the suspension volume effect on the luminescence. Using the same shaped bottles with different broth volume, we can see the effect of air-suspension interface area on the luminescence.

In our experiment, three test tubes containing *P. kishitanii* suspensions with different volumes (5, 10 and 20 mL) were used. Air-liquid interface areas ($S_{air\text{-}liquid}$) are the same for all three suspensions. The ratio of the interface area to the volume, $S_{air\text{-}liquid}/V$ differs for the three samples (Figure 5).

Oxygen is supplied to the bacteria only at the interface, because DO concentration inside the bacterial suspensions were almost zero. In a well-stirred condition, therefore, area for "breathing" per one bacterium is largest in the case of smallest volume, assuming that bacterial cell number is proportional to the suspension volume. The ratio of cells that suffer from this breathing problem to the whole cells would increase with volume, and more frequent fluctuation in luminescence would be observed. In fact, in a thermostated chamber luminescence from suspension with the least volume showed smallest luminescence fluctuation among the three (Figure 6).

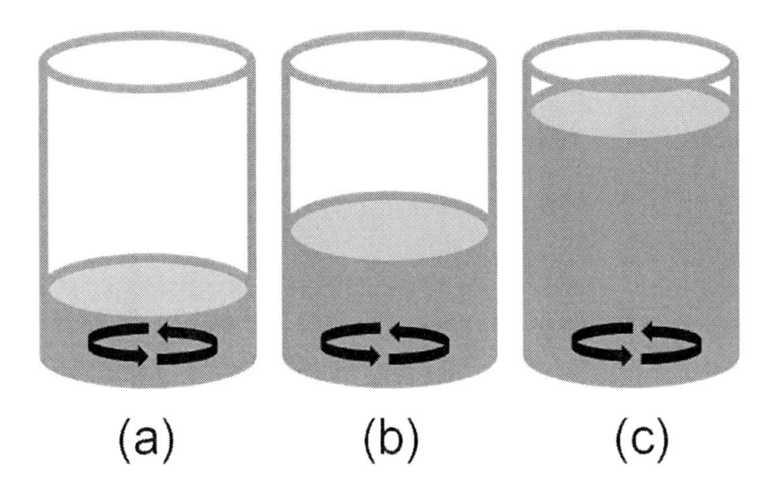

Figure 5. Suspensions of different volumes placed in test tubes. Suspension with smallest volume has the largest $S_{air\text{-}liquid}/V$ value (a), whereas the one with largest volume has the smallest value (c), respectively.

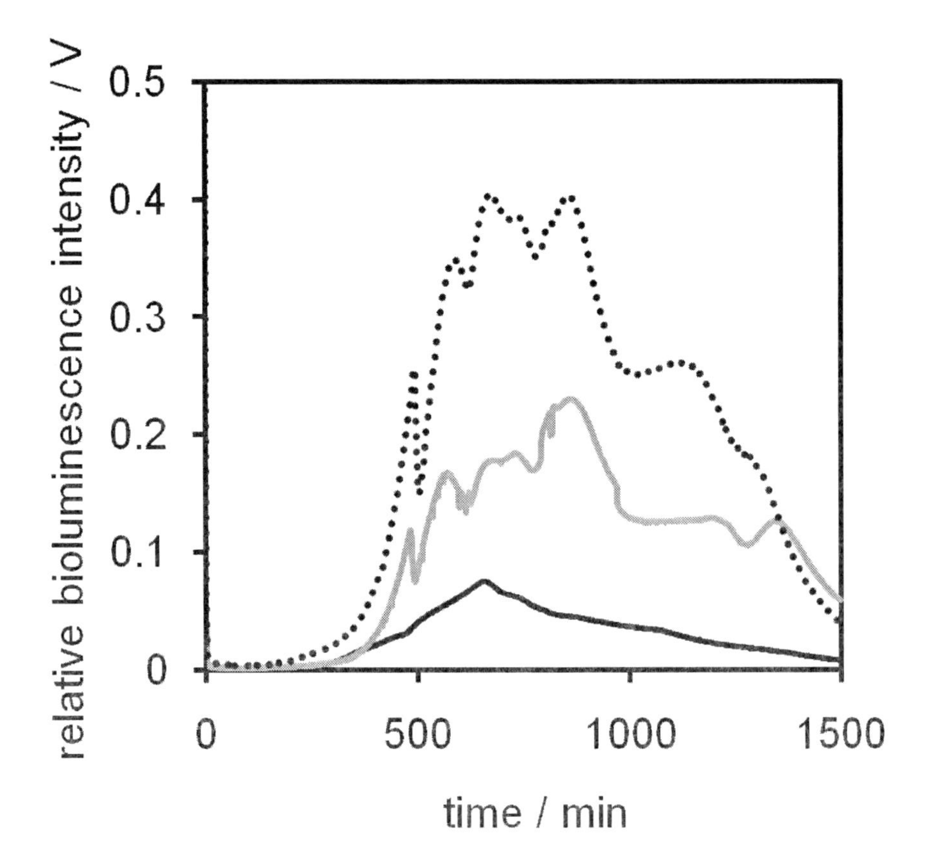

Figure 6. Bioluminescence of *P. kishitanii* suspensions with different volume; 5 mL (black solid line), 10 mL (gray solid line) and 20 mL (dotted line). Experimental conditions except for the volume were the same with the result shown in Figure 3.

In a project to fabricate a device using a stable luminescence from bioluminescent bacteria, species that show such fluctuations are not suitable. As far as *P. kishitanii* is concerned, controlling of the bacterial bioluminescence intensity by DO control is quite difficult, although on the other hand *V. fischeri* is well controlled to enable commercial application [19-21].

6. Control of Bacterial Luminescence

We learned that suspension of *P. kishitanii* temporally changes bioluminescent intensity even if we try to control DO. The intensity fluctuates spatially as well. We then tried to control the luminescence from the viewpoint of geometric symmetry in oxygen supply. We reported that *P. kishitanii* cells with larger motility show larger luminescence intensity [22] and that they tend to show oxygen taxis [15]. We can then imagine that bacterial cells located near the air-suspension interface are in the oxygen-rich environment. They would show different luminescent characteristics compared to those are located at the bottom of the bottle. What would happen if we use a bottle made of oxygen-transparent material? Polydimethylsiloxane (PDMS) is a material known to show a good oxygen transparency. We made a capsule with this material, put suspension of *P. kishitanii*, and sealed it with a PDMS cap. As a result a really stable bioluminescence was observed with spatial homogeneity (Fig. 7) [15].

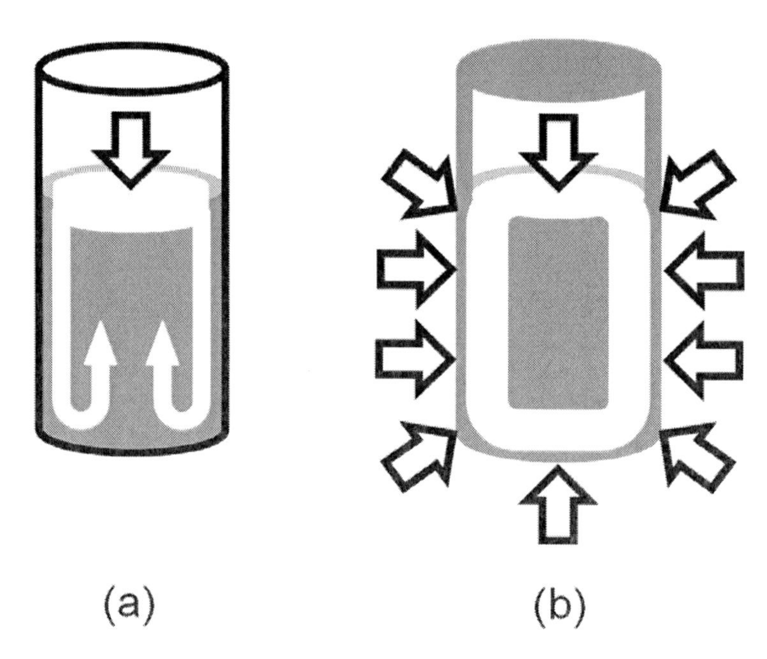

Figure 7. Schematic illustration of luminescence pattern in bacterial suspensions. In a glass cuvette, oxygen is supplied into the suspension at air-liquid interface, leading to the nonuniformity of DO distribution in the suspension, resulting in the convection (a). Using a capsule made of PDMS, oxygen penetrates from all the directions through the material into the suspension, allowing bacteria near the capsule wall to glow without convection (b).

CONCLUSION

Bacterial bioluminescence has been studied and is now applied in various fields. The bioluminescence varies with species, and in a certain condition it shows an interesting behavior such as oscillation. By using a suitable materials, oxygen dependence of the luminescence might be a viewpoint to develop a novel luminescence device in the future. As bioluminescence in the bacteria is the result of quorum sensing, bioluminescent bacteria can be a model organism for the study of other pathogens. The pathogenesity of *Vibrio cholerae*, which often attacks human in area after large natural disaster such as Tsunami, is controlled by the quorum sensing [23]. Prediction of when and where in our body the serious disease starts requires understanding of the bacterial behavior both from biochemical and nonlinear viewpoints. Study of bioluminescence in view of nonlinear science would be, in this sense, valuable for the future application of the knowledge to clinical field. Apart from such application, we can regard the bacteria as functional particles that communicate each other via an autoinducer. Study of the relationship between the function of each cells and the oscillation of the suspension should be a novel model of chaos and order in biological system. Further study of bioluminescent bacteria from various viewpoints are expected to have a whole illustration of the bacterial luminescence.

REFERENCES

[1] Haddock, S. H. D.; Moline, M. A.; Case, J. F. (2010). Biolumminescence in the sea. *Annu. Rev. Marine Sci.*, 2, 443-493.

[2] Ripp, S.; Daumer, K. A.; McKnight, T.; Levine, L. H.; Garland, J. L.; Simpson, M. L.; Sayler, G. S. (2003), Bioluminescent bioreporter integrated-circuit sensing of microbial volatile organic compounds. J. Ind. Microbiol. *Biotechnol.*, 30, 636–642.

[3] Ramanathan, S.; Shi, W.; Rosen, B. P.; Daunert, S.; (1997) Sensing antimonite and arsenite at the subattomole level with genetically engineered bioluminescent bacteria. *Anal. Chem.*, 69, 3380–3384.

[4] Hastings, J. W. (1975) Bioluminescence: from chemical bonds to photons. *Ciba Found Symp.* 31, 125-146.

[5] Lei, B.; Tu,S. C. (1998). Mechanism of reduced flavin transfer from *Vibrio harveyi* NADPH-FMN oxidoreductase to luciferase. *Biochemistry*, 37, 14623-14629.

[6] Watanabe H.; Mimura, N.; Takimoto, A.; Nakamura, T. (1975) Luminescence and Respiratory Activities of *Photobacterium phosphoreum* Competition for Cellular Reducing Power, *J. Biochemistry*, 77, 1147-1155.

[7] Karatani, H.; Yoshizawa, S; Hirayama, S. (2004) Oxygen triggering reversible modulation of Vibrio fischeri strain Y1 bioluminescence in vivo. *Photochem. Photobiol.*, 79, 120-125.

[8] Boettcher, K. J.; Ruby, E. G. (1995) Detection and quantification of *Vibrio fischeri* autoinducer from symbiotic squid light organs. *J. Bacteriol*, 177, 1053-1058.

[9] Miller, M. B.; Bassler, B. L. (2001) Quorum sensing in bacteria. *Annu Rev Microbiol,* 55, 165-199.

[10] Pe´rez, P. D.; Hagen, S. J. (2010) Heterogeneous response to a quorum-sensing signal in the luminescence of individual *Vibrio fischeri*. *PLoS ONE* 5(11) e15473.

[11] McIlvane, P.; Langerman, N. (1977) A Calorimetric investigation of the growth of the luminescent bacteria *Beneckea Harveyi* and *Photobacterium Leiognathi*, *Biophys. J.* 17, 17-25.

[12] Herring, P.; (2002) Marine microlights: the luminous marine bacteria, *Microbiology Today*, 29, 174-176.

[13] Strogatz, S. H. *Nonlinear Dynamics And Chaos: With Applications To Physics, Biology, Chemistry, And Engineering (Studies in Nonlinearity) . 1st edition.* Boulder, CO. Westview Press; 2001.

[14] Sato, Y.; Sasaki, S. (2006) Control of the biolumiescence starting time by inoculated cell density. *Anal. Sci.*, 22, 1237-1239.

[15] Sasaki, S.; Mori, Y.; Ogawa, M.; Funatsuka, S. (2010) Spatio-temporal control of bacterial-suspension luminescence using a PDMS cell. *J. Chem. Eng. Jpn.*, 43, 960-965.

[16] Sato, Y.; Sasaki, S. (2008) Observation of oscillation in bacterial lluminescence. *Anal. Sci.*, 24, 423-425.

[17] Karatani, H.; Heguri, A.; Hirayama, S.; Takeko Matsumura, I. T., (2007*) Space-Time Imaging of Bioluminescence from a Single Colony of Luminous Bacterium, BUNSEKI KAGAKU* 56, 43-46.

[18] Li, X.; Chow, D. C.; Tu, S. C. (2006) Thermodynamic Analysis of the Binding of Oxidized and Reduced FMN Cofactor to *Vibrio harveyi* NADPH-FMN Oxidoreductase FRP *Apoenzyme, Biochemistry*, 45, 14781-14787.

[19] Conforti, F.; Ioele, G..; Statti, G. A.; Marrelli, M.; Ragno, G.; Menichini, F., (2008) Antiproliferative activity against human tumor cell lines and toxicity test on Mediterranean dietary plants, *Food Chem. Toxicol.*, 46, 3325-3332.

[20] E., Fulladosa; E., Murat, J. C.; Martinez, M.; Villaescusa, I. (2005) Patterns of metals and arsenic poisoning in *Vibrio fischeri* bacteria, *Chemosphere*, 60, 43-48.

[21] Scheerer, S.; Gomez, F.; Lloyd, D. (2006) Bioluminescence of *Vibrio fischeri* in continuous culture: Optimal conditions for stability and intensity of photoemission, *J. Microbiological Methods* 67, 321-329.

[22] Sasaki, S.; Okamoto, T.; Fujii, T. (2009) Bioluminescence intensity difference observed in luminous bacteria groups with different motility. *Lett. Appl. Microbiol.* 48, 313-317.

[23] Higgins, D. A.; Pomianek, M. E.; Kraml, C. M.; Taylor, R. K.; Semmelhack, M. F.; Bassler, B. L. (2007) The major *Vibrio cholerae* autoinducer and its role in virulence factor production, *Nature*, 456, 883-886.

INDEX

#

A

B

C

D

E

F

G

K

L

M

N

O

P

Q

R

S

T

U

V

W

X

Y

Z